THE
GARDENER'S
PRACTICAL
BOTANY

THE
GARDENER'S
PRACTICAL
BOTANY

JOHN TAMPION BSc,PhD,MIBiol

DAVID & CHARLES : NEWTON ABBOT

ISBN 0 7153 5589 9

Set in 11/13 Plantin
and printed in Great Britain
by W J Holman Limited Dawlish
for David & Charles (Publishers) Limited
South Devon House   Newton Abbot   Devon

To my parents whose sacrifices made this work conceivable and my wife whose sacrifices made it an actuality.

Contents

		page
	List of Illustrations	11
	Foreword	13
ONE	HOW PLANTS GROW	14
	Leaves	16
	Roots	17
	The Vascular System	18
	Photosynthesis	21
	The Composition of a Plant	25
	Respiration	26
	Cells	27
	Hormones	29
TWO	TEMPERATURE, LIGHT AND WATER	34
	Temperature and Plant Growth	34
	Light	37
	Water Supply	40
	the flow of water in plants; leaves; roots; soil structure; soil water; the complete system	
THREE	MINERAL NUTRITION AND OTHER FACTORS AFFECTING GROWTH	52
	Nutrient Supplies from the Soil	53
	Organic Material in the Soil	58

Nutrient Elements—a Brief Survey   61
    nitrogen; phosphorus; potassium; sulphur;
    calcium; magnesium; iron; manganese,
    molybdenum, zinc, copper and cobalt;
    boron; sodium; fertilizers
Other Environmental Factors Affecting Growth   70
Pollution   71

FOUR   CONTROLLING PLANT GROWTH   75
Growth Habit   75
Geotropism   77
Cultural Practices   79
Chemical Treatments   82
Pruning   86
Overcrowding   88
The Size of Plants   89
Abnormal Growth   91

FIVE   SEEDS AND SEEDLINGS   93
The Formation of Seeds   93
Germination   96
Viability and Life Span   99
Dormancy   101
Conditions for Germination   106

SIX   VEGETATIVE PROPAGATION   110
Runners and Rhizomes   111
Bulbs and Corms   112
Tubers   113
Root and Stem Cuttings   119
Leaf Cuttings   123
Tissue Culture   124
Budding and Grafting   127

SEVEN   FLOWERING AND FRUITING   131
The Control of Flowering   133
Effects of Daylength   137

	Effects of Temperature	139
	Culture and Chemicals	141
	Pollination	144
	Fruit Set	146
	Fruit Growth	147
	Ripening	148
	Fruit Storage	150
EIGHT	**WEEDS, PESTS AND DISEASES**	152
	Weed Biology	153
	Weed Control	155
	burning; cultivation; competition; herbicides	
	Pest Biology	163
	Pest Control	164
	Disease Biology	167
	Disease Control	168
NINE	**GREENHOUSES, FRAMES AND CLOCHES**	173
	The Theory of Environmental Control	174
	The Structures Themselves	176
	Heating	180
	Watering	182
	Lighting	184
TEN	**BREEDING NEW PLANTS**	188
	Plant Names	189
	The Inheritance of Characters	190
	Chromosomes	193
	Mutations	197
	Breeding Techniques	198
	Formation of Polyploids	201
	Somatic Mutations	202
	Bibliography	205
	Index	213

9

List of Illustrations

PLATES

	page
Stomata in the leaf epidermis of *Tradescantia*	65
Cells and central vascular bundle in *Erica* leaf section	65
Bonfire smoke a potential hazard to health	66
A cultivated thistle, not a weed	66
Abscission layer forming at base of Virginia creeper leaf	83
A rose in need of pruning	83
Primordia in *Berberis* flower-bud section	84
Section through a flower of *Ribes* showing ovules	84
Section across the developing anthers of *Lilium*	117
Section of a single pollen sac of *Lilium*	117
Carpel from a buttercup flower	118
Section of wheat seed showing embryo	118
Fallen apples as sources of brown rot fungus infection	135
Part of a saprophytic fungus colony	135
Chromosomes of the broad bean prior to separation (*G. Holt*)	136
Chromosomes of the broad bean separating during cell division (*G. Holt*)	136

Drawings and Diagrams in the Text

Features of a typical dicot plant	15
The vascular system	19

Diagram of a plant cell 28
Etiolation 38
pH and availability of elements 55
Percentage composition of plant in terms of elements 62
Effect of pinching out 80
Development of plant embryo from fertilization to
    maturity 94
A peach seedling remains dwarf until it is chilled 103
Auxins stimulate the rooting of cuttings 122
Technique of apical meristem culture 126
Parts of the flower involved in seed production 132
Flowering response and daylength 134
Addition of gibberellins to a vegetable rosette 143
Bindweed reproduces both vegetatively and by seed 155
Weed seedlings 158
The spread of apple scab 170
Simple climatic orbits 174
Energy distribution in different forms of light 185
Dominant and recessive genes 192
Diagram to illustrate a plant-breeding experiment 200

Foreword

To many gardeners a botanist is merely someone who pulls the petals off flowers or indulges in incomprehensible arguments about the naming of plants. Nothing could be further from the truth.

Botanists are interested, just as gardeners are, in living plants. It seems futile, therefore, that many of the results of botanical research are locked up in books and periodicals that the gardener never sees. The present book has been worthwhile if it does no more than open the eyes of the average gardener to the possibilities of 'botany'.

In condensing so much work into so little a space I have necessarily made a personal selection of items for inclusion. To those gardeners whose problems are not even mentioned, let alone solved, I offer my apologies. My hope is that more gardeners will be inspired to delve into the literature of botany to find their own solutions.

My thanks are due to all those botanists whose researches have contributed to this book and to my wife whose typing has helped to create it.

JOHN TAMPION

Chapter One

How Plants Grow

Good gardeners are the men or women who understand their plants. Give identical cuttings or seeds to two people and one will produce healthy, vigorous plants while the other will have only shrivelled remains to show. In popular language this is due to the gift of 'green fingers', but if we look objectively at the problem the real difference is in the knowledge and skill of the gardener. To grow plants well we must be aware of the factors which affect plant growth. We must anticipate their reactions and know their individual requirements. Fortunately the task is not as formidable as it might at first seem because all plants have certain features in common. Learn these and the battle is half won.

Most garden plants can be put into one of three groups according to appearance alone. Firstly there are the trees, the shrubs and a few of the smaller plants which have woody stems. In the second group are the many plants which have distinct leaves and stems but are herbaceous rather than woody. They may be annuals, lasting only one year; or perennials which die back to the soil level each winter; or biennials, an intermediate group of herbaceous plants which over-winter once, often as a low rosette of leaves, and flower in their second year. Plants in these two groups almost all have two seed leaves and are known as *dicotyledons* or dicots for short. The third group is of plants that do not have readily visible stems except when flowering and the leaves normally arise from low down without any obvious stalks. The above-ground parts are rarely woody and the leaves are usually long and thin with parallel veins. These plants are *monocotyledons* (monocots) with only one seed leaf. To understand the way in which these different types of plant grow we must look at them in detail. For simplicity's sake we will leave out woody plants for

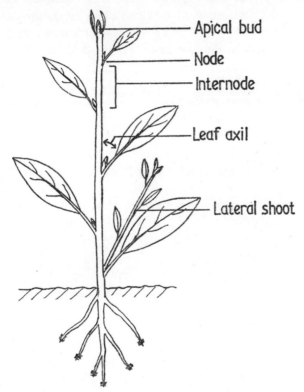

Apical bud

Node

Internode

Leaf axil

Lateral shoot

Five important features of a typical dicot plant

15

the moment, but their special features are discussed in later chapters.

Just as the different organs of an animal each have their own function, so do the different parts of a plant. The important regions of a dicot are illustrated on page 15.

LEAVES

The leaves are the organs which build up many of the complex substances needed for plant growth from much simpler ones. This role has little influence upon the shape of a leaf except to restrict its thickness so that gases can easily reach the inside; its final shape depends on how long its various parts go on growing. Leaves first start to develop in vegetative buds and never grow out haphazardly from a length of stem. Buds also protect the delicate growing points of the stems themselves. Magnified under the microscope the bud at the tip of a plant (the *apical bud*) has a cone-shaped appearance, and on this cone small lumps are formed which grow progressively larger as they get further away from the tip. These tiny lumps, known to botanists as leaf *primordia*, represent the beginnings of the individual leaves and each one eventually grows into a mature leaf. They arise in a very regular way; usually in a spiral when the mature leaves also spiral around the stem, but in plants such as those of the mint family which have leaves in opposite pairs the primordia are also formed opposite one another.

As a leaf primordium grows, it soon becomes flattened. If the whole length of its margin goes on growing at approximately the same rate the edge of the leaf will eventually become gently curved from apex to base, producing an *entire* leaf, such as that of green privet. If certain parts of the margin slow down or stop growing the edge will become wavy, like an oak leaf, or even complexly dissected, as in a carrot leaf. The way in which the growth rate of a leaf primordium changes is usually a characteristic passed on from the previous generation, and many plants can be identified by the shape and size of their leaves. The stalk (*petiole*) of a leaf is clearly different in structure from the blade and shows its own particular pattern of growth, often quite distinct from that of the leaf blade. The leaves of dicots usually stop growing when they have reached a certain size and they may remain at this mature size for weeks, months or even years, before they finally die and fall off. Every leaf has a region of the stem associated with it and in the angle (*axil*) between the upper side of the leaf and the stem an axillary bud is present. Often this remains dormant, but if conditions become suitable it may develop into a new branch at some later date.

The region where a leaf arises is called a node. The length of the stem depends mostly on the amount of growth occurring in the internodes between the leaves. In a biennial rosette this growth is very small, while in a climbing plant the internode may be several inches long; usually it is between these extremes, allowing air to circulate freely round the leaves and a reasonable amount of light to fall upon each of them. In monocots, the stem often elongates only during flowering. The increase in height of the plant is due to the continuing growth of the base of the leaves, producing new leaf rather like a tape from a machine. The dying-off of the tip of a monocot leaf is therefore of only minor importance to the plant. Dicot leaves, with their more limited capacity for growth, are usually seriously affected by such damage, particularly if it occurs when they are young.

ROOTS

Below the ground, in the roots, a different organization exists. The growing root apex is not protected in a bud but instead has a *root cap* which it pushes before it through the soil. The root cap produces slimy material which probably helps to protect the delicate apex from mechanical damage or drying up. The pres-

17

ence of primordia sticking out at the tip would obviously hamper its growth through the soil and so the *lateral* (side) roots grow out a long way back from the root tip, in the region which is no longer growing in length. Instead of arising as mounds on the surface of the main root these lateral roots actually form inside it and then push through to the outside. This again may help to protect them since they are strongly growing by the time they emerge.

Damage to the main root apex, which frequently occurs, has a relatively small effect on these lateral roots. One of the functions of a root is to take in water from the soil. To assist in this task many roots develop *root hairs,* which are tiny outgrowths of the cells of the root surface arising just behind the elongating region of the root. They penetrate between the soil particles, sticking firmly to them. If a plant is pulled roughly out of soil, these root hairs are torn off and the root is less able to take in water to supply the needs of the shoot.

THE VASCULAR SYSTEM

The various parts of a plant are linked together by the vascular system, through which water and food materials can flow round the plant. Actually it is a double system. One part, the *xylem*, is mainly concerned with the flow of water and simple mineral substances taken from the soil. The other, the *phloem*, is involved in the movement of more complex 'foods', hormones and so on, through the plant. The vascular system extends completely throughout the plant, as shown opposite. A skeleton leaf shows some of the finer branches very clearly.

A plant is made up of tiny units called cells and each cell consists at first of a tiny blob of living matter surrounded by the cell wall. Some cells in the xylem are very well suited for their task because they are elongated, just like tubes, and the living contents, which would hamper the flow of water, are dissolved away

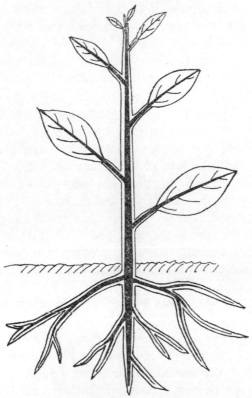

The vascular system, penetrating into every part of the plant, forms a complete circulating system for water and nutrients

as soon as the cell reaches its full size. All that remains are the dead walls, with their characteristic thickenings that help to prevent the collapse of the tubes when they are placed under stress by the flow of water through them. Other cells also occur in xylem in addition to the tubular ones. The flow of water in the xylem is almost entirely from the roots to the leaves.

The phloem also contains minute tubes, placed end to end. They are not empty, however, but contain special living contents different from any others in the plant. The sections are linked together through the large holes in their end walls, which are known as sieve plates. One of the main substances flowing through the phloem is sugar—identical, in fact, to the ordinary sugar we use in our food and drink. Despite years of research botanists still do not know exactly how the sugar is pumped through the phloem. It is known, however, that the flow may be in any direction and is controlled by the growth taking place in the plant. Growing or storage regions act as 'sinks' into which the nutrients drain. The rate of movement can easily be found by weighing, say, a developing pumpkin every day; of course a great deal of the weight is due to water but the increase in solid matter can be calculated. Virtually all this is formed from substances produced in the rest of the plant and flowing down the narrow stalk into the pumpkin. Remembering that the flow takes place only in the phloem of the stalk, which is a small part of its cross-sectional area, and that the sugar may flow in a 20 per cent solution, a rate of flow of about 63 inches per hour can be calculated as a minimum, possibly much higher in places because of the presence of barriers across the tubes, such as the sieve plates.

The composition of the liquid in the phloem has been determined by using that familiar garden pest, the aphid. Aphids have long needle-like sucking mouths which they carefully insert into the phloem tubes and suck out the sugary juice. Usually they take more than they need and we get drops of sticky 'honey-dew' all over the leaves. Botanists have killed aphids as they suck the phloem sap and severed the mouth tube from the rest of the aphid. If the end of the tube is prevented from sealing up, the sap from the plant will go on being pumped out for some time and enough can be collected for chemical analysis.

The term *sap* is used in gardeners' language to describe any liquid which flows in a plant, but botanists distinguish between the phloem sap and the xylem sap. When we talk about the sap rising in the spring it is the xylem sap we mean. At that time of the year the food reserves formed in the previous year and stored over the winter are mobilized and passed up to the growing buds. The solution which flows contains much less sugar than the phloem sap but in some plants, like the birch tree, it is strong enough to be tapped off for wine-making! Later in the season, when the leaves have developed, the xylem sap contains hardly any sugar and usually does not ooze out very much when the stem is cut.

PHOTOSYNTHESIS

This sugar which flows from one place to another in the plant is an *organic* material. This means that it contains the element carbon and has been produced by a living organism. Inorganic substances, by contrast, are present as components of the soil and air and provide the raw materials for the production of organic matter. To remain alive all living organisms need a supply of energy. The sun is the ultimate source of this energy, which reaches the earth in the form of light and heat. Plants are able to convert light energy into the energy (calories) contained in organic substances by the process known as photosynthesis. Life on earth is entirely dependent upon this energy conversion, so it is not surprising that botanists have spent a lot of effort in un-ravelling the mysteries of the process. It is so complex that the full details are not yet known but during the last twenty years great advances have been made in our understanding by using the sophisticated modern apparatus and technology of bio-chemistry and biophysics.

It is easy to show that in the light many green leaves produce starch. Starch turns black when treated with ordinary medicinal tincture of iodine, but before testing a leaf the green colour, which would tend to obscure the reaction, is normally removed from the leaves by immersing them for a short time in methylated spirit. An interesting experiment which clearly demonstrates that light is needed for the production of starch can be carried out by placing a black paper cut-out stencil on both sides of a leaf for a few days. If the leaf is then removed and tested for starch this is found only in the parts that have been exposed to the light.

Ordinary sugar dissolves in water easily at room temperature and a plant would have difficulty in storing reserves of sugar because these would be continually moving around the plant with the flow of water. To overcome this problem the plant links the small individual sugar 'units' together to form starch and aggregates the starch into large grains. Starch grains do not easily dissolve in cold water and the plant is continually building up starch as a food store and splitting it down into the sugar units again so that it can be moved about from one part of the plant to another. The potato provides an excellent example of this continuing process. When we plant a seed potato it is full of starch, but as the eyes grow out the starch is converted into sugars and used by the growing shoots until eventually the shrivelled old tuber will contain hardly any starch at all. As the leaves develop they produce starch which is then split down to sugars and moved about the plant. Some is used up to provide new top growth; some is deposited again as starch in the stem, close to the vascular tissue. Most of the sugar is built up again as starch in the developing new potatoes, where it forms the reserves for further growth in the next season. All this can be easily demonstrated with the iodine test.

The real question, however, is where does the starch come from in the first place? Starch and sugar are examples of the group of substances known as *carbohydrates* which means that they consist of the elements carbon (symbol C), hydrogen (H) and oxygen (O). By using a special radioactive form of carbon, it can be shown that the carbon in starch comes from the carbon dioxide gas (CO_2) present in the air. During photosynthesis CO_2 gas is taken in by the plant and oxygen gas (O_2) is liberated. It is this oxygen which animals need to breathe in order to survive, and in fact green plants in the dark and all non-green parts of plants also need oxygen. Scientists think that all the oxygen gas of the atmosphere has been produced by plants, over the millions of years that they have existed on the earth. The oxygen liberated during photosynthesis comes from the water (H_2O) taken in by the plant. The plant joins the hydrogen from the water to the carbon dioxide from the air by a series of separate steps, eventually producing sugar and then starch. The energy needed to carry out the steps comes from sunlight and the stages are very complex.

The light energy is first absorbed by the green *chlorophyll* pigments and most of it is eventually changed into the chemical energy of something known, for short, as ATP. This energy-rich substance is capable of passing on its energy to bring about, for example, the *fixation* of carbon dioxide gas. The splitting of the water, which was previously mentioned, is also brought about by the use of light energy.

The colours of the rainbow remind us that ordinary 'white' sunlight is made up of waves with different energy content, visible to man as colours. In fact the colours shade into one another and there is a continuous range from ultraviolet rays, which are not visible to the human eye, through the visible colours of violet, indigo, blue, green, yellow, orange and red to the infrared rays which are again invisible to the human eye. The energy content of light becomes progressively less per photon (a 'package' of light energy) from the ultraviolet to the infrared region. Only certain colours contain photons of the correct energy for photosynthesis. If 'white light' is shone through chlorophyll, either in solution in a tube or in a leaf, the blue and red components will be preferentially absorbed, so that what emerges is mostly the remaining colour—green.

Artificial light usually contains a different distribution of energy between the various wavelengths and does not always give the same growth responses as natural light. This point is returned to in later chapters (see especially pages 184-7).

Over the millions of years that plants have existed on the earth they have become adapted to particular growing conditions. Thus some plants prefer open sunny situations and are capable of benefiting from the full amount of sunlight, while other plants are more suitable for shaded places where the available light intensity is much lower. These are the shade plants and they may well be damaged by an excess of light energy if planted in open sunlit areas; the sun plants, on the other hand, may not be able to obtain enough energy for their normal growth if they are planted in shady spots. Some garden plants are very demanding in the conditions they require but fortunately most are quite tolerant and are able to exist in a range of positions. This is partly because plants can respond to the amount of light available by changes in their growth. The leaves on the sunny side of a plant, for instance, are usually smaller and thicker than the 'shade leaves' on the poorly illuminated side. These tend to be thinner and have a larger surface area.

The green chlorophylls are clearly essential for photosynthesis. How do we then explain the existence of plants with yellow leaves? It can be shown in the laboratory by suitable extractions and identifications of the pigments that yellow leaves also contain chlorophylls but the amount may be only one-tenth or even less of that contained in a normal green leaf. It is common knowledge that yellow-leaved ('golden') cultivars are less well able to manufacture their own food and under natural conditions are at a disadvantage by comparison with their green counterparts. In the garden we can take special care of them, reducing competition from other plants or cutting out any normal green branches which develop on them so that the golden form can continue to flourish. Frequently plant breeders opt for a halfway position in which the leaves are variegated, with yellow margins or blotches. Plants which truly lack chlorophyll—albino plants

—are not capable of existing once they have used up the food reserves stored in the seed, so that though we occasionally see them in the seed bed they never reach maturity. 'Copper' and other dark coloured leaves also contain chlorophylls but the green colour is masked by other pigments which are not involved in the process of photosynthesis. This is easily shown by boiling a leaf in water. The darker colours dissolve into the water leaving a green leaf—the chlorophylls do not pass so easily into the water. This only applies to leaves taken during the normal growing season; autumn colours are a different matter and usually replace rather than supplement the green chlorophylls.

THE COMPOSITION OF A PLANT

So far, only the way in which carbohydrates are formed and move about the plant has been discussed. Although much of the structure of a plant consists of carbohydrates other constituents are also present, such as proteins and fats. Proteins are more complex than carbohydrates and contain other elements in addition to carbon, hydrogen and oxygen. Nitrogen (N) is one of the most important of these as it forms part of all the amino-acids, the simple building blocks from which proteins are constructed. Although the air contains a great deal of nitrogen gas this cannot be incorporated into amino-acids by plants unless it is first 'fixed' by a micro-organism. Only certain plants, particularly those of the Pea family, which have small *nodules* (lumps) on their roots, have a close association with the correct micro-organism. Other plants obtain their nitrogen in inorganic forms from the soil, the commonest being ammonium and nitrate. These both dissolve very easily in any water in the soil and can easily enter roots. Nitrogen is needed in large amounts by plants and is therefore called a major or *macro-nutrient*.

For normal plant growth other major nutrients are also needed. Phosphorous (P) is particularly involved in the transfer of energy from one substance to another within the plant. Potassium (K) is present in all cell sap and calcium (Ca) in the cell walls. Chlorophyll contains magnesium (Mg), while some amino-acids have sulphur (S) in them, and both of these elements are sometimes included with the other macro-nutrients.

Even if supplied with all the elements already mentioned a plant will still not grow normally unless small amounts of minor (micro-) nutrients (also known as trace elements) are available. Most of these are associated with some component of the plant which is essential to its life but which only needs to be present in very small amounts. These elements are iron (Fe), boron (B), copper (Cu), manganese (Mn), molybdenum (Mo), zinc (Zn) and sometimes chlorine (Cl) and perhaps sodium (Na). The macro- and micro-nutrients come almost entirely from the soil and details of their supply will be given in Chapter 3. For the moment we need only to note that no plant will grow to its full potential unless it is supplied with the correct amount of all these essential inorganic nutrients.

RESPIRATION

The way in which light energy is converted into the chemical energy contained in sugars has already been described. The problem which now remains is how the sugar is converted into the other constituents of the plant such as the fats, proteins, other nitrogen-containing substances, pigments and the hundreds of other components which are essential to its life. The conversions are made by way of thousands of individual steps, each quite small and controlled by the presence of a special type of protein known as an *enzyme*. Each step has its own particular enzyme. If a particular enzyme is missing then the plant is unable to carry out that particular step, and if it is one that is absolutely necessary for life the plant will not be able to grow and will die. Some steps, such as the production of coloured pigments in the flowers, are not so essential and the plant will continue to grow and reproduce regardless of whether or not it contains the enzyme for that step.

So the sugar is broken down, remodelled and rebuilt into other components. Even these may not be entirely stable and may be broken down again and converted into something else. These dynamic processes of building up and breaking down form the basis of life and growth. If they slow down, growth will be slow. If we can stimulate them, growth will be rapid. Many of the conversions need energy and a special set of reactions exists which breaks down sugar into simpler substances and produces energy for interconversions at the same time. This complex process is called respiration and is obviously totally different from the familiar human process of breathing in and out. It is sometimes called cellular respiration to make the difference clear. All living organisms, both plant and animal, carry out at least some cellular respiration. It can be considered, though this is simplifying it very much, to be a reversal of photosynthesis because the overall result is a conversion of sugar into carbon dioxide and water. Oxygen is used up in this process. One of the important energy-transfer substances produced by respiration is ATP, which was mentioned previously in connection with photosynthesis. The P stands for phosphorus, which is part of the ATP molecule, emphasizing that element's vital role in growth.

CELLS

All plants are made up of separate microscopic units called cells, which exist in a wide variety of sizes and shapes, many of them specialized to perform particular roles in the plant. The elongated conducting cells of the xylem and phloem have already been mentioned. Photosynthesis is carried on in special cells; the plant surface is covered with another type of cell, and so on. In some regions of the plant there are cells whose main function is

27

to divide to produce more cells; such cells are called *meristematic* and they occur especially at the stem and root tips, in the buds. They also occur in the cambium of woody plants (see Chapter 6). Apical meristems are the regions containing dividing cells at the tips of stems and roots. Once produced, new cells can grow and develop into the many types needed to give a complete plant. The position and surroundings of a young cell largely determine the type of mature cell into which it will develop. Under the microscope the different types can be recognized by their shape,

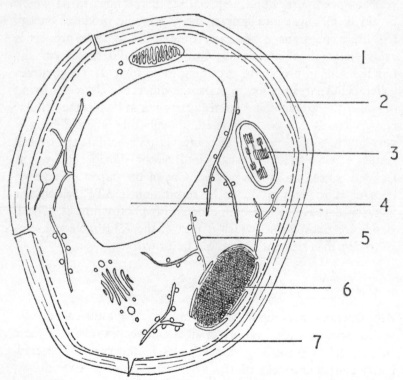

Diagram of a plant cell showing (1) site of respiration, (2) cell wall, (3) site of photosynthesis, (4) cell sap, (5) site of protein synthesis, (6) nucleus, (7) outer cell membrane

the thickness and markings of the cell walls which surround them and by their contents.

The living processes which we have been discussing do not all occur together in one jumbled bag of chemicals. The more we magnify a cell the more fine detail and structure we can see. The reactions of photosynthesis, for example, occur in rather ovoid structures called *chloroplasts*. Cellular respiration mostly occurs in even smaller particles known as *mitochondria*. Completely beyond the limits of the light microscope and only visible with the enormous magnification obtained with the electron microscope are the *ribosomes* in which protein synthesis occurs. These and some other *organelles* (literally 'tiny organs') of a typical cell are shown opposite.

All this complexity requires some means of control, just as a factory needs a manager to keep production at the required rate and prevent over- or under-production of some vital component. In the cell it is the nucleus which is chiefly responsible for the control of cell activity. It contains nearly all the information required to keep the cell functioning, and can cause changes in a cell's activities by switching on or shutting off the cell's supply of the particular enzymes needed to bring about a given reaction. The nucleus is discussed more fully in Chapter 10 because of its relevance to plant breeding.

HORMONES

An individual cell does not live in isolation. The surrounding cells produce chemical 'messengers' (*hormones*) which can enter adjacent cells or be carried in the conducting systems throughout the plant. These superimpose their control on that of the cell's own nucleus—rather like the area manager! Even this is not the full story because the plant is growing in a particular place and is subject to environmental influences which can affect its growth and development. Usually the responses to these influences also involve the hormones.

The chemistry of plant hormones is totally different from that of animal hormones. Plant hormones and the many synthetic hormone-like substances which have been produced by chemists are usually grouped according to their effects. Those which have been known for the longest period of time and have therefore been most extensively studied are known as the *auxins*. The main natural auxin occurring in plants is known as IAA, an abbreviation of its chemical name. Auxins can cause cells to grow longer and this can be demonstrated in the laboratory by a special test. A small cylinder is cut from the sheath covering the first true leaf of a young oat seedling which has been grown in the dark and when this cylinder is treated with the correct amount of an auxin it increases in length. Any chemical causing such growth is said to possess auxin activity. IAA has many other effects on plant growth which will be mentioned in several of the later chapters. Often the effects of synthetic auxins are not identical in every respect to those of IAA and they may even be more useful to the gardener because of this. The effect of auxins depends upon the amount used: too much may easily inhibit or distort growth. This fact has led to the development of many hormone weed-killers.

A second group of hormones is also involved in the control of elongation. These are the *gibberellins*. Originally they were identified as substances produced by a fungus disease of rice which caused a great elongation of the rice stem, and a similar elongation in other plants when applied to them. It was only many years later that botanists were able to prove that gibberellins are naturally present in plants and involved in normal growth. The delay was caused because gibberellins occur in very minute amounts and they are difficult to identify because of their very complex structure. Over two dozen different gibberellins have now been found, all of them with the same basic structure but differing in the finer details of the molecule. Only a few synthetic chemicals have been produced which have gibberellin-like activity, presumably because of the complexity of the natural substance. The different gibberellins are designated in a series A_1, A_2, etc. The common form available is gibberellic acid (GA_3), produced by controlled growth of the fungus which caused the original disease of rice. Tests for gibberellin activity are more complicated than those for auxin activity. One is based on the ability of gibberellins to cause the production of the enzyme which breaks down starch into sugar in the food reserves of the barley seed. The various gibberellins often show subtly different effects in the other tests. Some weedkillers and growth-retardants work by interfering with the natural production or role of gibberellins in plants.

The *cytokinins* are a third group of natural hormones. The limits of this group are much less well defined than are those of the auxins or gibberellins. Cytokinins seem particularly concerned with the divisions of cells rather than with their elongation. This is not a clear-cut role because the application of auxins, and in certain cases gibberellins, can also cause cells to divide, as in the use of auxins for the rootings of cuttings. Any part of a plant containing a high concentration of cytokinins appears to act as a sink which drains nutrients from other parts of the plant. This may explain the way in which developing fruits and seeds, which contain high levels of cytokinins, are able to draw on the reserves contained in leaves and other parts of a plant to support their further growth. This may also be the way in which the products of photosynthesis in the leaves are channelled to different parts of the plant.

Apart from these three well-known groups of hormones there are many other natural substances which affect plant growth. Amongst these is the recently discovered *abscisic acid*. This, as its name implies, is involved in the *abscission* (falling-off) of leaves. At one time it was thought that leaf-fall was due to changes in the auxin concentration but now it is known that abscisic acid is also involved. Abscisic acid also promotes dormancy in plants and its effects in this respect are mentioned again in later chapters.

Several other hormones have been postulated as accounting for various aspects of plant growth, but until they are isolated and identified they remain a topic of argument among botanists.

So far the different hormones have been considered in isolation from one another. In a living plant, of course, there is an ever-changing interaction between them, and the final growth of the plant depends on this. By manipulating hormones we can obtain particular growth responses and many gardening practices rely upon this for their effectiveness, even though we may not realize exactly what we are doing. Reference will frequently be made to hormones in the other chapters of this book.

SOME GROWTH SUBSTANCES
AND THEIR MAIN EFFECTS

Effect	*Auxins*	*Notes*
Promotion of cell elongation		At low concentrations involved in normal stem growth
Inhibition of cell elongation		At higher concentrations used as weedkillers (see page 161)
Curved growth		Involved in phototropism and geotropism (see pages 39 and 77)
Promotion of cell enlargement		Involved in the swelling of fruits and storage organs (see page 147)
Promotion of cell division		Involved in regeneration and cambial growth (see Chapter 6)
Stimulation of root formation		Involved in the rooting of cuttings (see page 119)
Inhibition of lateral bud development		Important in apical dominance (see Chapter 4)
Inhibition of the abscission of organs		Involved in leaf and fruit fall (see page 148)
Production of parthenocarpic fruits		For example, in pineapple (see page 146)

Gibberellins

Promotion of stem elongation	Especially of intact plants. Causes bolting of rosettes, increases the length of internodes and converts dwarf plants into tall ones (see Chapter 4)
Promotion of cell division	Especially in the sub-apical meristem (see page 90)
Breaking of dormancy	For example, in buds of birch and seeds of lettuce (see page 103)
Production of parthenocarpic fruits	For example, in pears (see page 146)
Promotion of flowering of rosettes	For example, replaces requirement for chilling in carrot (see page 142)
Promotion of enzyme production	For example, during germination of cereal seeds (see page 31)

Cytokinins

Promotion of cell division	Demonstrated in tissue culture experiments (see page 125)
Promotion of cell enlargement	For example, root cells of tobacco may be larger and stems thicker than normal
Promotion of organ formation	For example, increases the number of buds on leaf cuttings of *Sainpaulia* (see page 124)
Inhibition of root formation	But roots may be thicker
Counteraction of apical dominance	Stimulates lateral buds to grow out (see page 82
Delay in senescence	Of part of plant containing the cytokinin

Abscisic Acid

Reversion of the effects of other hormones	Depending on species and part of plant used will reverse the effects of auxins, gibberellins and cytokinins
Promotion of dormancy	Found in some seeds and in some leaves before abscission

c

Chapter Two

Temperature, Light and Water

In this chapter we discuss the effects of three external influences upon plants. These are just some of the factors that will determine whether a plant reaches its full potential in our particular garden. A good gardener finds out what the conditions are in all parts of his garden, and then selects his plants to suit them. No garden is ideal for all plants but with a little knowledge great improvements can be made in the range of plants grown and in the quality of the fruit or flowers produced.

TEMPERATURE AND PLANT GROWTH

One environmental factor which we are all familiar with is temperature. In spring we long for the end of frosts and the rising temperatures needed for plant growth; indeed it is frost damage that the gardener usually thinks of first when temperature is mentioned. There are upper temperature limits for plant growth but fortunately they are rarely reached in temperate climates, except in greenhouses. Each particular species of plant, and even differ-

34

ent cultivars of a single species may have its own particular temperature requirements for optimum growth. For practical purposes the gardener usually works with much rougher limits. To call a plant 'hardy' is, more or less, synonymous with saying that it will stand frost during the winter. Half-hardy indicates that it will be killed by frost.

Why do some plants live on after frost while others fail to? Botanists are not completely sure of the exact details but several general points can be made. Remember first that the plant is made up of cells. These are bounded by cell walls, and contain the various delicate organelles mentioned previously (see page 28). Frost damage is usually considered to be due to strains and stresses caused by water changing from its normal liquid state into ice crystals, in or around the cells and back to water again when the plant thaws. Experiments have shown that the damage caused is less if we can keep the strains and stresses to a minimum. If your plants have slowly frozen during the night they may not yet be killed, so there is real value in going out early to protect them from early morning sun which could well cause a rapid thawing and consequent death, or at the very least a slowing of normal growth.

Even plants which are 'hardy' are found to be able to stand different amounts of cooling at different times of the year, and many plants can be made more resistant to frost by 'hardening'. This is a slow process of acclimatization to gradually lowering temperature and usually occurs naturally during the onset of winter in a wide range of plants—from cabbages to trees.

Some plants, such as begonias and sunflowers, are just not capable of being hardened against frost. Presumably there is something in the working of their cells which prevents the necessary changes taking place. In plants which do harden there is usually a reduction in the amount of starch present and an increase in the amount of sugar. It is possible that the starch grains of unhardened plants cause extra disruption during frost damage.

35

There is usually a difference in temperature between the air and the soil. The surface of bare soil warms up rapidly in the morning as sunlight reaches it and cools as the strength of the sunlight diminishes. The temperature of the soil depends upon the absorption and retention of the sun's heat and the gardener can exercise a small amount of control over this. For example, because a dark soil absorbs more heat than a light-coloured one he can apply soot to obtain a slightly warmer soil. At night, cool air accumulates around the plants and settles on the soil, and because air is a poor conductor of heat the soil is prevented from passing its heat to the cooler plant above. On clear nights particularly the plants radiate out their heat and their temperatures can drop sufficiently to cause frost damage even though the temperature of the soil is not at freezing point. A cover of vegetation over the soil has the effect of reducing both the daily and the seasonal variations because of its own absorption and radiation of heat. A mulch may have the same effect as vegetation. In experiments on strawberry beds the temperature below a straw mulch was over 6°C (13°F) higher than that over the mulch. A marked lowering of air temperature occurred above the mulch. Obviously strawberries or any other plants protected from frost by straw should be well covered and not allowed to poke out through the straw. Snow itself, of course, forms a protective cover over soil and reduces the loss of heat from it. Fogs, either natural or artificial, help in preventing air frosts caused by rapid heat loss from air around plants.

Soil temperature is of great importance to a plant because if the soil is too cold absorption of water and minerals by the roots will be very restricted, and even if the air warms up during the day the plant will not be able to grow normally. But we should not imagine that it is only very low temperatures which are bad for plant growth. It is also possible to have too high a temperature. As the temperature of a plant rises the rate at which its living processes take place increases. It can be roughly estimated that for every 10°C (18°F) increase in temperature there will be a doubling in the rate of life processes. This increase continues until the delicate balance between the various reactions is upset. Just as the proteins in an egg white are coagulated by heating so are the protein enzymes in the plant damaged by heat. Some plants can withstand higher temperatures than others before growth is affected and, as might be expected, such plants usually originate from countries with warm climates. The effects of high temperatures are complicated by the fact that not all the reactions in a plant are equally affected by a rise in temperature. The building up of food reserves by photosynthesis is usually damaged by heat more quickly than the breaking-down reactions of respiration.

Finding an optimum temperature for plant growth is further complicated by the fact that even with a single species of plant the best temperature for growth may vary with its stage of development. Seedlings often need a higher temperature than more mature plants, which fortunately fits well with the general practice of germinating seeds in a warm place and then gradually transferring them to rather cooler conditions outside. Certain plants have even turned their optimum growth to this daily temperature variation between night and day. Tomatoes, for example, prefer a lower temperature at night than during the day.

LIGHT

Light is a second major environmental factor affecting plant growth. The role of light in photosynthesis has already been discussed in Chapter 1 and we shall be concentrating here on other influences of light. One of the most obvious effects of low light intensity, both indoors and out, is the excessive elongation which is produced. The extreme occurs in plants grown in total darkness. Provided such a plant already has a store of food, it will go on growing for some weeks but its appearance will be completely abnormal. If we put germinating seeds away in a dark cupboard or let potatoes sprout in the dark we soon produce this effect. In the garden itself this extreme type of damage is usually only seen when we have carelessly put a light-proof box or cover down and forgotten it.

The extreme condition is probably the best one to examine first. For the particular effects which are about to be described the term *etiolation* is used, but in practice the manifestation of etiolation may range from extreme to very slight depending on the amount of light which is available to the plant. Mild etiolation due to poor light is often seen in the garden.

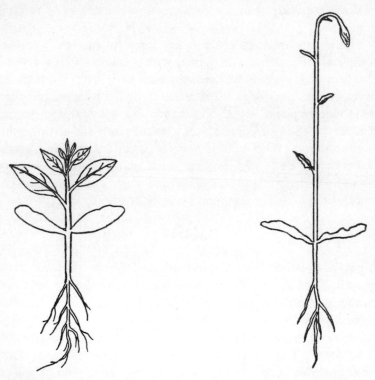

Light-grown and dark-grown seedlings, to illustrate etiolation

One of the most obvious signs of etiolation is that the plant is pale yellow or white rather than green. This is because the last stage of chlorophyll synthesis which gives the green colour cannot take place without light. Another obvious effect is that the stem becomes greatly elongated (page 38). This is because the internodes between the leaves grow much longer than is normal and not because more leaves are produced. This, and other changes in the cells of the stem, cause the etiolated plant to be weak and spindly. Once this elongation has occurred it is impossible to reverse because cell growth has taken place and there is no way in which the cells can be made smaller again.

Gardeners take advantage of the effects of etiolation in certain plants. With rhubarb, celery, chicory and so on we can 'blanch' the plants by covering them with a bucket, roll of paper or something similar. The rapid elongation produced is of great benefit and the reduced green colour and failure of the leaves to expand properly is not a disadvantage. Of course we cannot go on blanching a plant continuously throughout the season otherwise it would not be able to build up its reserves by photosynthesis and would exhaust itself.

Many plants respond to light coming from one particular direction by growing towards it. We have all seen this response in boxes of seedlings on the greenhouse bench or potted plants on a window-sill. It is known as phototropism, and is due to greater growth of the stem on the shaded side of the plant. Experiments have shown that there is more of the auxin IAA on the shaded side, causing greater elongation of the cells there and the consequent turning of the stem towards the light. Blue light is particularly effective in causing this response.

The length of daylight varies from day to day, the amount of variation depending on the latitude. In temperate countries there is a marked seasonal difference, with long summer days and short winter ones. It is not surprising that many plants have linked particular stages of their growth to such variations. The flowering of some plants is controlled in this way and will be discussed in more detail in Chapter 7. The induction of resting buds in the autumn and sometimes their 'breaking' in the spring is also often controlled by daylength.

WATER SUPPLY

Attention must now be given to a third important environmental factor—the supply of water to the plant. Without water there can be no life because all the reactions which go on in growing plants only do so in the presence of water. Dry seeds are living entities but will not grow until water has entered them. This is understandable if it is remembered that 90 per cent or more of a growing plant is water. Until photosynthesis starts a young seedling is actually using up the solid matter in it, and its increase in size is due to the uptake of water plus a certain amount of re-organisation of its component molecules.

The Flow of Water in Plants

The soil acts as a reservoir of water which a living plant can draw upon to replace water which has evaporated from its surface. The water-conducting xylem in the plant (already described in Chapter 1) forms the major path along which water flows from the soil to the leaves.

When water movement in a plant is discussed the botanist has two basic concepts in mind which must now be explained. Firstly the plant consists of cells which are joined together. Lining the walls of living cells are delicate membranes which control the entry to and exit from the cells of sugars, the inorganic minerals and all the other molecules necessary for life. All the cell walls, dead cells and the spaces between the cells can be considered as one complex interconnected network through which water and dissolved substances can flow easily from one place to another in the plant. This is known to botanists as the 'free space' of the plant. If a chemical or mineral in solution is supplied to the roots of a plant it rapidly enters the free space and can spread all over the plant.

Movement of water in or out through the cell membrane is controlled by the relative strength of the solutions on the inside and the outside of the membrane. If the inside solution contains more dissolved substances than the outside, as it normally does, water accumulates inside.

The net flow of water from a dilute to a strong solution, if these are separated by a membrane, is known as *osmosis*. Normally there is a gradient of osmotic forces in a plant which assists water movement from the roots to the leaves and growing points. The movement of water by osmotic forces is quite slow and does not account for the rapid movement of water which occurs in a plant. This rapid movement is the result of evaporation of water from the plant into the air around it, causing a tension to be set up through the xylem of the plant and down into the roots. Water is drawn into the xylem from the free space, carrying with it dissolved minerals from the soil and other substances produced inside the plant itself.

The amount of water lost from the exposed surfaces of a plant is dependent upon the structure of the plant. The entire surface of a young plant is covered by a layer of cells known as the epidermis. The outer walls of the epidermal cells on the shoot have a waterproof material deposited in and over them, differing in thickness from plant to plant. Plants growing in regions where they have difficulty in maintaining their normal water content (this may be due to a variety of causes) usually have a thick waterproof covering, often supplemented by additional waxy deposits. The photosynthetic parts of a plant have pores in the epidermis through which can pass the gases used and produced during photosynthesis and respiration and also, incidentally, through which water can evaporate into the outside air. These pores are called stomata (a single one is a stoma) and they are constructed

41

so that they can be opened or closed according to the prevailing conditions (see page 65). This control is extremely complex in its details but generally speaking it can be assumed that in a normal plant the stomata will be open in the light, when of course photosynthesis is occurring, provided the water supply is adequate. If too much water is being lost by evaporation from them, the stomata close until the loss is made up by absorption through the roots. When the stomata are closed photosynthesis is restricted because the supply of carbon dioxide (CO_2) from the outside air is reduced. The stomata respond quickly to changes in both water and carbon dioxide supply.

In fact there is always a lag between loss of water from the shoots and absorption by the roots so that plants contain less water during the day and make good the deficit at night. On hot sunny days they may even show signs of wilting during the day, only to recover at night.

Although a continuous spray of water over plants can be very useful in dry weather it is a mistake to go out during full sun and soak the leaves with the hose. Certainly the cool water will temporarily raise the humidity around the shoots and reduce water loss but it will also upset the delicate stomatal mechanism which has adjusted itself to the sunny dry conditions, so that when you stop supplying water adverse effects may occur. A second problem is that droplets of water may act as magnifying glasses through which the sun's rays will 'burn' the leaf surface. Furthermore, if the water is 'hard', minerals will be deposited on the leaf surface as the water evaporates and these also may upset the osmotic forces and draw water out of the leaf cells. We shall return to the subject of watering a little later.

Older parts of plants normally grow a protective cork layer which replaces the epidermis. The exact form of the cork differs according to the plant but the basic principles of its formation are the same. A layer of cells starts to divide and produce thick-walled cells which are water- and gas-proof and give excellent

42

protection. If the *cork cambium* (as the dividing cells are called) forms a continuous ring round the stem the cork will be produced in rings, although it may crack later as it is pushed outwards by new cork forming inside. Scaly bark develops when successive, separate, cork cambia are produced and separation occurs along the lines of weakness. *Lenticels* in the cork, where the cells are loosened and powdery, allow gases to pass in and out. The visual appearance of lenticels on the outside of the bark is usually elongated because they increase in size as the tree grows older and larger.

Leaves

The actual area of leaf, stem and flower exposed to the air is obviously an important factor in the water loss of a plant; the greater the surface area the more water will evaporate. Plants which grow in regions where the water supply is poor often show a reduction in surface area. Thus leaves may be very small or remain on the plant for only a short time. The leaf-fall of deciduous trees is usually thought to be a means of reducing water loss when water, although plentiful in the soil, cannot easily get into the plant because of low soil temperatures. Other plants may have no leaves at all and instead have green stems to carry out photosynthesis. These may even become flattened and rather leaf-like but will still retain the structure of a stem. Further reduction of the surface area can be obtained by massing all the plants into one spherical or cylindrical structure. This is the shape we associate with many cacti and succulents and it is interesting that similar shapes can be reached in quite distinct plant families. Certain members of the cactus family, for example, closely resemble some members of the spurge family.

The structure of a leaf is not quite as simple as we have so far imagined it. In a typical leaf from a dicot plant we can distinguish an upper (*adaxial*—towards the leaf axil) and a lower (*abaxial*) surface. Inside the leaf this difference is reinforced by a distinction within the internal tissues—the leaf *mesophyll*. This is separated into fairly closely packed *palisade* cells in the upper part and much looser *spongy mesophyll* in the lower half. The photosynthetic reactions occur in both types of mesophyll cell. Because of the air spaces between the cells the surface area exposed to the air by cells inside the leaf is much higher than the total external surface area of the leaf. The internal air spaces are connected to the outside air by the stomatal pores. According to the species of plant stomata may occur on both sides of the leaf or only on the lower surface. See photograph on page 65.

In most monocot leaves the two halves of the leaf are similar in structure and there is no distinction into palisade and spongy mesophyll. This probably reflects the fact that the linear leaves of monocots are held more or less erect and do not have such an obvious visual separation into upper and lower surfaces.

Roots

The remaining part of the water conduction system which has not been considered so far is the root. Although old roots can take in a small amount of water it is mostly absorbed through the younger portions of roots. Here again surface area, this time of the roots, is of importance. The delicate root hairs provide one means of increasing the surface area, the existence of plenty of small growing roots another. One of the major reasons for the loss of plants during transplanting is that the parts of the roots most able to take in water are also the ones which are most easily lost when a plant is dug up. Hence the great importance of digging plants out with a good ball of soil and of not disturbing the roots of a pot-grown plant too much when planting out of doors.

Water from the soil enters the roots as a result of the osmotic forces which have already been described. The cell sap of the root hairs is more concentrated than the soil solution outside and hence water is drawn into the root hairs. If we put too much soluble fertilizer on the soil, or the land is flooded by salt water, the soil solution may become stronger than that in the roots and water will not be able to pass readily into the roots. Naturally this state of affairs is followed by the wilting of the plant and, if prolonged, by death. Death may also be speeded by a direct toxic effect of the high concentration of minerals. Since young growing roots are involved in water uptake it follows that soil conditions need to be suitable for root growth and extension. Roots not only need a supply of the products of photosynthesis from the shoots above ground and a supply of minerals and water from soil but also a supply of oxygen so that they can carry out the energy-releasing reactions of respiration. This may be in short supply in the soil if conditions are not favourable or the soil structure is not satisfactory.

Soil Structure

Soil structure is of vital importance to soil fertility and scientists have therefore made an intensive study of it. All soil is derived from one or more of the parent rocks of which the earth is made up. Over millions of years, by the action of wind, rain, frost, sun and other natural forces, the parent rock is weathered into small particles. These may be carried from one place to another so that the soil in any one region may not always be derived from the rocks beneath it. There are many different types of rock and even more types of soil. The subject is complex enough for a lifetime's study so all that can be offered here is a very simplified picture.

The first stage in looking at a soil is to examine the mineral particle size. By convention the particles in a soil are placed in the following categories, according to size:

		Approximate size in inches
Gravel	— above 2mm	above 8/100
Coarse sand	— between 2.00 and 0.20mm	8/100 to 8/1000
Fine sand	— „ 0.20 and 0.02mm	8/1000 to 8/10,000
Silt	— „ 0.02 and 0.002mm	8/10,000 to 8/100,000
Clay	— below 0.002mm	below 8/100,000

The proportion of these different particles in a soil has an important bearing on its properties and can be easily determined by taking a dry sample of soil and passing it through sieves of known mesh size. Another, simpler, method is merely to shake some of the soil in water and let it settle out in a tall, narrow glass vessel. The larger the particles the more quickly they settle. When this is completed the amount of each particle type can be roughly estimated. Obviously large gravel or stones should be removed first and estimated separately if necessary but as far as plants are concerned the important thing is the relative proportions of sand, silt and clay in the soil.

The minute clay particles are often so small that they do not settle out from water but instead form a *colloidal suspension,* that is one in which the particles are too large to be truly dissolved but too small to settle out. (Starch also gives a colloidal 'solution'.) The particles in water-saturated clay can flow over one another, which is why it is slippery, and why houses need special foundations if built on pure clay. In damp clay the particles are held apart by thin films of water, but when clay dries out the water evaporates and the particles move closer together. Hence the clay shrinks and cracks appear. One way of preventing this aggregation of clay particles in garden soil is by the addition of substances such as calcium, which flocculates ('deposits') the particles, or, better still, plenty of organic material to weld the minute clay particles into larger granules which have the beneficial properties of clay with respect to plant nutrients but do not compact together. Some organic material is more effective than others and there are a number of commercial products available which will keep clay in workable condition for as long as a year.

The terms used to describe soil texture are combinations of the three primary components together with the term 'loam' which indicates a reasonable proportion of all three. So there can be silty clay loam, sandy loam and so on. Once you know the percentage of the three components in your particular garden soil you can discover its accepted textual name by reference to the appropriate diagrams in almost any book on soil.

Apart from inorganic minerals, any soil which will support plant growth usually also contains organic material, the dead remains of plants or animals incorporated into the soil. This will be in various states of decay from raw organic, through compost to well-degraded humus. The organic material helps to bind the mineral particles into larger granules and so to provide a good crumb structure. The crumbs, which under normal cultivation do not break down into their components, allow good aeration of the soil because of the large air spaces between them.

Soil Water

Once the solid matter has been accounted for it is the spaces between the particles that make up the rest of the bulk of the soil. These may be full of air or water but their size is determined by the size of the solid particles. If the particles are large then the spaces between them will also be large; smaller particles fill in these gaps and between them smaller ones still. A remaining feature of particle size is that the smaller the particle size the greater will be the surface area when compared, weight for weight, with larger particles.

How does soil texture affect water supply to the plant? When rain falls onto a dry soil it soaks in and fills up the spaces between the particles, pushing the air out as it does so. If there is sufficient rain the soil will become completely saturated and even the large gaps will be full of water. In this condition the soil is not suitable for plant growth because it will contain no oxygen. When the rain stops the excess water will start to drain out from the larger spaces between the particles and be replaced by air. The speed at which this drainage occurs will depend upon the underlying material and drainage. If it is chalky or sandy then rapid draining will occur but if there is a layer of clay this will hold back the water and slow down the soaking away process.

The lie of the land is also important here and improvements in drainage are often possible by inserting drainage channels. To be successful such schemes need to be well planned, the nature of the subsoils fully examined and drainage points located where they will not cause trouble by waterlogging some other part of your own or a neighbour's garden. Since drainage schemes involve hard work and expense they should never be undertaken without proper planning. Excessive run-off from land can also cause trouble by carrying away the soluble nutrients of the soil and should not be encouraged. A steady, gentle, draining away is usually what is needed. Most gardeners prefer to work with their natural soil-drainage conditions and select plants to suit it rather than try to modify it. As with all gardening practice it is better to work alongside nature than to try and meet it in a headlong clash!

But, to return to our theoretical soil, once the excess water has drained away the soil reaches a condition known as *field capacity*. It is holding as much water as it can. The amount depends upon the soil texture and composition. Sand cannot hold more than about 10 per cent, loam takes 20 to 30 per cent and clay even more. As the water evaporates the forces holding the water to the particles get stronger and the plant finds it increasingly difficult to absorb water from the soil. Eventually the limit of *available water* is reached, the plant is unable to take any into its roots and consequently wilts. The soil is then said to be at the *permanent wilting point* but the actual amount of water remaining again depends on the type of soil: for sand it may be about 3 per cent, for loam between 10 and 15 per cent and for clay over 20 per cent. On the whole the type of plant does not make much difference to the numerical value of the water content at the wilting point—mostly it is the soil itself which determines this.

It should now be obvious why it is of little use to merely damp the surface of dry soil. The water rapidly soaks in, spreads out over the surface of the dry soil particles and so becomes unavail-

able. Prolonged watering of this type encourages roots to grow
into the surface layers and if watering is stopped for any reason
the roots are particularly susceptible to drying out. Watering
must be carried out thoroughly to ensure the provision of adequate available water to a reasonable depth. There are certain
advantages in having a surface layer which is loose. The gardener
usually achieves this by hoeing between the plants. Soils possess
the ability to conduct water from one place to another because
water can flow as a continuous film over the soil particles. If the
soil is loose and the particles are a long way apart the film of
water will break up into separate parts and the water will not be
able to flow. This prevents water being drawn from the deeper
soil to replace that which is lost by evaporation from the surface.
A second beneficial effect is to allow any rain which does fall to
penetrate the soil and not run off. The value of hoeing to control
weeds is related to the drying out of the surface as well as the
actual damage to the weed seedlings. If the soil remained damp
many of the weeds would develop new roots and recover. The
main disadvantage of hoeing is the risk of damage to shallow
roots of the garden plants themselves.

Hoeing to give an inch or so of dry soil could be considered as
a 'dry soil' mulch, though the gardener normally thinks of a layer
of straw, leaves, peat, or partly decayed plant material when the
word mulch is mentioned. From the point of view of water
supply the mulch acts as a block to evaporation from the top
layers of the soil and particularly protects roots near the surface.

The covering of ground with black plastic is also probably
best described as mulching because the effect is similar. The
crop—say potatoes—is planted to come up through small holes
in the sheet. Rain runs into these holes and is retained under the
sheet. Because light does not penetrate the plastic any weeds
which germinate will be etiolated and finally die because the
holes in the plastic will be filled by the shoots of the crop. One
problem might be a local shortage of oxygen in the soil but since

the crop comes through the holes it is not likely to suffer.

The Complete System

The separate parts of the water picture must now be pieced together. Water evaporates from the leaves and shoots into the air because the air is usually not fully saturated with water. The warmer the air is the more water it can carry. Moving air also increases the rate of evaporation because it prevents the building up of layers of moister air over the plant surface.

In the plant, water moves from roots to leaves, drawn by the tension set up by water evaporation from the surface of the leaf cells. Its passage is mainly in the xylem and, to a lesser extent, the rest of the free space of the plant.

It is easy to demonstrate that it is normally tension and not pressure which effects the movement of the water. When a stem is cut off air is drawn into it, but if a pressure had existed liquid would be pushed out, and this only happens under the special circumstances noted below. The water loss from the plant seems to be an unavoidable consequence of the need to have pores through which carbon dioxide and oxygen can pass in and out of the plant. The flow of water does however have some beneficial effects, probably because minerals and other simple substances are moved rapidly through the plant in the *transpiration stream* (transpiration being the botanical term used to describe water evaporation from the plant). The water which evaporates from a plant is, of course, pure ('distilled') water and all the substances that were being moved about in the water inside the plant are left behind. If the water supplied to the roots of a plant contains too great a concentration of minerals the excess may be deposited in or on the leaves, causing damage.

Another aspect of water movement which must be considered is *root pressure*. The roots of some plants, at certain times of the year and under certain conditions, can push water upwards into the shoots. This is usually demonstrated by cutting off a shoot and attaching a device for measuring pressure to the cut stump. The forces which cause this are thought to be osmotic. The more concentrated sap in the xylem draws water from the soil into the plant, causing a pressure to develop. This can be quite high, sufficient to push water to the top of a medium-sized plant, but probably not to the top of a tall tree. It may be important in getting the water moving early in the year when the leaves with their large evaporating surfaces have not yet developed, and under humid conditions when evaporation is restricted. The sap which flows from cut trees in the spring is forced out by root pressure. The sugars in this xylem sap come from food reserves stored in cells over the winter and mobilized again in the spring.

Chapter Three

Mineral Nutrition and Other Factors Affecting Growth

So far the soil has been considered as an inert medium from which the plant obtains water; but this is clearly a simplification. As we have seen in Chapter 1, plants need a supply of a wide range of elements in order to grow correctly and now we must look at the chemistry of soil particles and at the soil solution, which is another term for the water in the soil. The elements needed by plants have already been listed, but before going any further we must know how much of each is needed. Botanists have solved this problem by growing plants in solutions of known chemical composition instead of in soil. By using pure chemicals the effects of deficiencies and excesses of individual elements can be observed and the best proportions of elements for a given plant worked out. For good plant growth the correct balance of nutrients is needed; too much or too little of any one can have disastrous results. The gardener, then, has to find out what elements are available in the soil and to supply any which are lacking.

NUTRIENT SUPPLIES FROM THE SOIL

Soil is very variable and some growers have dispensed with it altogether and grow their plants entirely in carefully controlled nutrient solutions. Although of great interest and possessing some advantages over normal growth in soil, this soil-less culture (also known as *hydroponics*) is not usually able to compete economically with traditional methods. Great care is needed in looking after nutrient supplies in soil-less culture, ordinary soil acts as a buffer against sudden changes in the composition of the soil solution. In water culture it is also essential to aerate the water, otherwise roots go short of oxygen (which does not dissolve well in water) and growth is poor. This problem can be overcome by using an inert support such as gravel or sand which can be periodically flooded with a nutrient solution and then drained to allow aeration. The details of hydroponics are too complex to discuss here but the interested reader should consult other books on this subject.

Soil is a very complex mixture of different substances. A fertile soil contains not only the various elements needed for plant growth but also many which are not essential, together with a wide range of micro-organisms which are able to change one substance into another. Although it can be done, there would be little point in a chemist taking a sample of the soil, breaking it down into its individual components and finding out how much of each element is present. This would give a poor indication of the value of the soil for growing plants because not all the elements in a soil are available to the plant. Only those which can dissolve and enter the soil solution are available. The rest mainly act either as an inert support material or form a reservoir of elements which can be slowly released into the soil solution. Fertilizers are also divided into two groups—fast acting soluble ones which rapidly enter the soil solution and slow acting ones which liberate the nutrient elements slowly over a long period of time.

Too much nutrient dissolved in the soil solution upsets its osmotic strength and hampers movement into the roots. Another problem with excess application of soluble fertilizer is that it may be washed out of the surface layers of the soil by rain before the plants can absorb it, and if it finds its way into streams and rivers it can cause pollution and death of the fish. This may seem rather surprising but happens in the following chain of natural events. The water weeds are stimulated by the fertilizer and grow luxuriously (at the farmers' expense!). When the weeds die they start to decay. Decay is caused by micro-organisms in the water which use up the dissolved oxygen and sometimes even liberate noxious substances into the water. The fish which depend on the dissolved oxygen in the water to 'breathe', are deprived of their necessary supplies and die.

Soluble nutrients can also be lost from the soil solution by conversion into an insoluble form, often by chemical combination with some other component in the soil. Every gardener knows that there are acid soils and alkaline soils and that some plants can be grown in one and not in the other. The terms acid and alkaline are taken from chemistry and chemists use the 'pH' scale to describe the extent of acidity or alkalinity of a solution. The neutral midpoint is pH7. As the scale is descended towards pH1 the solution becomes more and more acid. In the opposite direction, up towards pH14, the solution becomes increasingly alkaline. The pH value of the soil is of great importance in controlling the availability of nutrients.

The extremes are never reached by soil solutions, which usually fall somewhere in the range from pH4 to pH8. Significant differences occur in the availability of different elements within even this range. This is illustrated opposite. Taken overall the maximum number of different elements is available in the narrow range between pH6.5 to pH7.0 and most plants do well if the soil pH is in this region. The farther the pH value is from this range the more restricted will be the list of plants which can

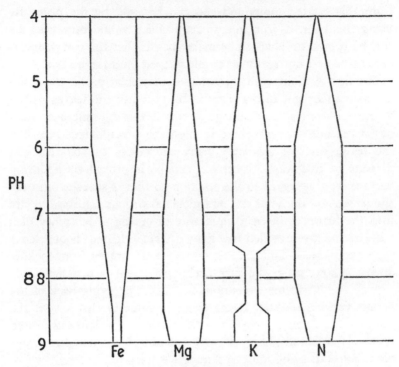

The effect of pH upon the availability of four essential elements

be grown. The plants which can grow at the pH extremes, on acid peaty soils or alkaline chalky ones are perhaps better described as plants which can tolerate low availability of certain elements rather than as plants which love acid or chalk soils. A common example of low availability is provided by the element iron. No matter how much iron is added to a chalky soil it will not be available to the plant because as soon as it touches the soil it is deposited out of solution. One way to prevent this is by *chelating* the iron, which simply means that the iron is protected from being precipitated by another chemical with which it combines loosely. When it enters the plant the iron is freed from this carrier compound and can be used by the plant. By using this form of iron many plants which would otherwise die can be grown on chalk. The main snag is that the compound is expensive and not really economic if used on a large scale.

Another way in which nutrients, particularly the trace elements, can be got into a plant is by spraying the leaves with a dilute solution. The advantage of this is that the micro-nutrient never reaches the soil where it might be precipitated. Provided the leaves are not too waxy many substances can be absorbed directly in this way. Absorption can be improved by adding a surface wetting agent to the solution so that it spreads out over the leaves and does not run off quickly as droplets. Entry directly into the leaves is often a rapid way of curing a deficiency of a trace element—provided it is known which element is deficient! The experienced botanist can often guess, simply from looking at the symptoms in the plant, which element is deficient. Particular growth forms, colours in the leaves or dying back of the shoots can reveal this. If the soil pH value is also known the diagnosis is more certain. Even the ordinary gardener can often make a reasonable diagnosis with the aid of the detailed symptoms listed in some mineral nutrition books.

The determination of soil pH is very easy. A botanist would use a pH-meter which gives a direct reading on a dial when the probe is immersed in a soil solution—usually an equal mixture of soil and distilled water is used. The gardener can buy a soil testing kit containing a chemical indicator which changes colour according to the pH value. By comparing the colour with the standard colour scale provided, the pH value can be determined with sufficient accuracy for all normal gardening purposes. This only takes a few minutes to do. Even in a single garden pH values may differ from one spot to another so it is wise to take the pH value where you are actually going to plant. The soil in some spots may be better for one plant than for another. Over the years scientists have worked out the best soil pH values for many important crop plants and these can be found in books dealing with plants and the soil, but many garden plants have never been fully investigated in this way. The intelligent gardener uses his own observation: if a plant does badly year after year but the nutrient supply seems perfectly adequate for other plants and no sign of disease can be found, then it may be the soil pH which is at fault.

It is usually quite difficult to change the pH value of a soil because most soils possess buffering action. A *buffer,* to the chemist, is a substance which tends to maintain its original pH value, even when acid or alkali is added to it. Of course a temporary change can easily be made but if the additive is soluble in water it is soon washed out and the soil returns to its original pH value. With continued cropping or the persistent additions of sulphur-containing fertilizers for years on end the pH value may drift lower and reduce yield. Lime and other calcium-containing soil dressings may cause a rise in soil pH if used persistently, and be equally bad.

To return to the soil particles themselves, we can now look at their contribution to the nutrient supplies in soil. Stones and gravel contribute virtually nothing in the way of nutrients and act mainly by encouraging drainage. Sand is mostly silica, SiO_2, but the silicon (Si) is not needed by plants and the oxygen (O) is not available to them. It would not be true to say that the element silicon does not occur in plants because some plants, particularly grasses, contain silica crystals which appear to contribute to the strength of the tissues. Maize, for example, can contain almost as much silicon as it does nitrogen. The difference is that the plant can grow without the silicon, but will not grow if the nitrogen is missing.

From the nutritional viewpoint it is the small inorganic minerals, particularly the clays, which are of great significance. Clays themselves are very complex substances containing various elements. The major ones are silicon (Si), aluminium (Al) and oxygen (O). Many minor components also occur including such important plant nutrients as iron (Fe), magnesium (Mg), zinc (Zn) and manganese (Mn). Many other elements are associated with the surface of clay particles and this surface area is very large. The result, in simple terms, is that clays act as a reservoir for many of the elements needed by plants, releasing them according to the prevailing conditions. In this way calcium and potassium, which might otherwise be leached (washed) out easily from a soil are held, and remain available for plants.

ORGANIC MATERIAL IN THE SOIL

Before the individual nutrient elements are discussed some consideration must be given to organic material in the soil. When plants or animals die they fall onto the soil and the complex molecules in them are broken down again into simpler substances. The breakdown is brought about by the action of various types of soil organisms. Worms and other larger soil animals may start the process, which is then taken over by the enormous numbers of micro-organisms that occur in fertile soil. The nutrients in organic matter only become available to plants when the complex substances have been completely broken down. Prior to this the beneficial effects of organic matter are largely due to its influence upon general soil structure and properties such as water retention. Over 90 per cent of the solid matter in a plant consists of the three elements, carbon, oxygen and hydrogen in the form of organic nutrients such as sugars, fats or proteins. The soil micro-organisms use these to provide energy for their growth, liberating carbon dioxide and water as waste products. A rich organic soil gives off quite appreciable amounts of carbon dioxide. Nitrogen and the inorganic elements remain in the soil to be re-used.

The process of decay and the liberation of nutrients is speeded up if the plant is first consumed by a large animal and converted into manure before it reaches the soil. Although the animal absorbs some of the nutrients from the plants the physical form of the remainder is changed so that it decays more rapidly when it does reach the soil. Raw vegetable matter and some fresh manures are not normally applied directly to the soil around growing plants for several reasons. Firstly, substances which inhibit plant growth are often present and these need to be broken down or washed away by rain. Secondly, the micro-organisms which break down organic matter in the soil would compete with the plant roots for oxygen and certain inorganic elements which are not plentiful in the organic matter itself. A deficiency of nitrogen, for example, often follows the application of raw vegetable matter to the soil. Considerable debate still rages over the respective merits of direct application and composting before application, but to the ordinary gardener matters of convenience often decide which course is to be followed.

A compost heap is designed to allow micro-organisms to break down the organic matter before it is put onto the soil. This turns the coarse vegetable matter into a friable, water-holding material. The heap must be properly constructed to obtain the maximum benefit but the principles are simply stated. The plant material must be damp but not saturated, and the heap should be firm but not so tight that air cannot get in. Often a shallow trench is taken out and the plant material put in, in layers 5 or 6 inches deep, with a sprinkling of nitrogenous fertilizer or a little manure on top of each. The addition of a little lime helps to prevent the heap from becoming too acid, but should be kept apart from the fertilizer. Alternatively a proprietary compost 'accelerator' can be added according to the instructions on the packet. A gentle damping with a watering-can may also be necessary. Each layer is separated from the next by a thin layer of soil which helps to hold the nutrients in the heap. If this is not done nitrogen may be lost as ammonia gas during the composting. The heap is then earthed over the top. It is sometimes more convenient to build up the heap in a container or frame which can be opened up and removed when the heap is large enough, although a 'lid' of polythene or other waterproof material should be left over the top. If correctly made the heap should start to warm up quite soon and the temperature rise, due to the heat produced by micro-organisms as they break down the vegetable matter, should be sufficient to kill any weed seeds.

One reason for covering the heap is to keep the heat in. Another is to prevent rain washing out the valuable soluble substances formed during rotting—unless you want to grow your best plants on the site of the compost heap! Too much water in the heap will exclude air and slow down the process of decay. Woody material will not decay quickly enough and should be burnt, together with diseased material and any plants recently treated with herbicides. Unless the compost heap can be properly made, it is safer to burn any seedling weeds or underground parts that might regenerate, otherwise they will be spread about the garden with the partly rotted vegetable mater. The heap can be ready for use in as little as two months during the summer but takes longer in the winter because the micro-organisms do not grow so rapidly at low temperatures.

The ash from bonfires is another 'natural' source of nutrients which can be added to the soil. During burning the complex substances built up by plants are quickly broken down to carbon dioxide gas and water. These are both lost into the air and the ash consists of the inorganic elements which originally came from the soil. Actually a little nutrient is carried away as tiny particles in the smoke. Bonfire smoke contains traces of chemicals known to cause cancer and the amount of these chemicals produced by a slow fire can be several hundred times that found in tobacco smoke. It is obviously advisable to reduce bonfire smoke to a minimum and this is easily done by making sure that the plant material is dry before lighting and ensuring a good supply of air to the bottom of the fire by the use of an incinerator. With plenty of air the conversion is complete to carbon dioxide and water and the intermediate toxic substances are not produced. (See photograph on page 66.) Ash is traditionally a good source of potassium but will, of course, contain all the elements needed by plants almost in the proportions present in the original plant material. Not all will be in an available form so that some weathering in the soil may be needed before they are liberated.

NUTRIENT ELEMENTS – A BRIEF SURVEY

Nitrogen

This element is essential for all plant growth but is particularly important for the formation of leaves and stems. Its effect on growth is magnified because the larger leaves produced when adequate nitrogen is present are also capable of a greater amount of photosynthesis than the small, stunted leaves which develop when it is scarce. The young leaves of nitrogen-deficient plants are pale, but as they grow older they develop bright yellow, orange or red colours and die off prematurely, with a consequent loss in yield from the affected plants. In the much rarer case of too much nitrogen in the soil, plants seem to produce more new cells, with a high protein content, and less of the carbohydrates which give strength to the cell walls. The result is lush, sappy growth which is rather fragile and easily damaged by droughts, frosts, pests and diseases. Much of the growth is in the form of leaves, and reserves in root crops may not be as large as expected because of this.

Nitrogen is removed from the soil when a crop is harvested. It can also be lost by leaching away in heavy rainfall and by conversion into volatile gaseous forms by certain soil microorganisms. This latter process is known as denitrification. Nitrogen is returned to the soil in organic materials, especially of animal origin. Nitrates and ammonium salts are common inorg-

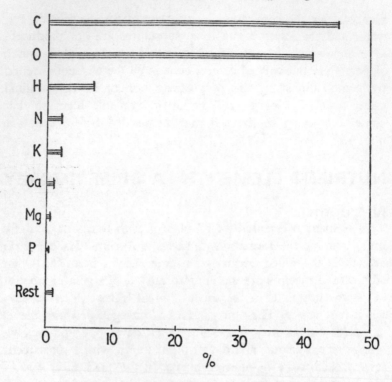

The percentage composition of a plant in terms of the individual elements

anic forms of nitrogen which the gardener can add to his soil. Comparisons between fertilizers are usually made in terms of the amount of the element N which is present, rather than by the actual weight of fertilizer added to the soil, because different percentages of nitrogen may be present. Certain micro-organisms can convert the nitrogen gas of the air into inorganic nitrogen in the soil. Legumes and one or two other plants have formed a partnership with micro-organisms and 'fix' nitrogen gas in root nodules for use in their growth.

Phosphorus

Phosphorus deficiency resembles nitrogen deficiency in producing stunted growth, smaller leaves and premature defoliation. Usually only a little available phosphorus is present in soil at any one time. When a soluble form such as superphosphate is added, much of its becomes unavailable quite rapidly, which also means that little of it is leached out of the soil. Applications of dilute solutions or thinly spread powdered forms react more readily with the soil and are therefore less efficient means of supplying phosphorus than localised application of larger granules. Maturity and ripening are hastened by phosphorus but although an excess of phosphorus, particularly on dry soils, may cause this to happen so quickly that yield is slightly reduced such problems are unlikely to occur in the average garden. Apart from the rapidly acting superphosphates and various ammonium phosphates there are several slower-acting forms such as basic slag and bonemeal.

Potassium

Particularly involved in the production of seeds, fruits and root crops, potassium promotes hard growth which is disease and frost resistant. If a soluble form of potassium is added to a soil, much of it will be washed out readily by subsequent rains. Clay particles can hold a certain amount of potassium in an available form but sand will not hold any. Sandy soils are therefore more likely to show the signs of potassium deficiency. Typical symptoms are scorching of the edges of leaves, stunting of growth and poor yields. Problems caused by excess potassium are due to an upsetting of the uptake of the other essential elements from the soil. Wood ash was one of the first forms of potassium fertilizer but other soluble forms such as sulphate of potash and potassium chloride are now frequently used.

Sulphur

Under normal circumstances sulphur deficiencies are rarely seen because the element is added as a component of nitrogenous or phosphatic fertilizers. But recent trends towards 'high analysis' N and P fertilizers as alternatives to the traditional ammonium sulphate and superphosphate (which contains sulphur as a valuable impurity) have meant that sulphur deficiencies may soon become more prevalent. The reduction of air pollution has also reduced the amount of sulphur returned to the soil; formerly sulphur dioxide gas given off during combustion used to be returned to the soil dissolved in rain, at least near centres of population. When deficiency does occur it usually results in yellowing foliage and rather thin, stiff woody growth. Orange and red tints may develop on foliage. Sulphur may be added (as noted above) as a component of fertilizers used mainly to supply some other element, and the addition of elemental S (the yellow 'flowers of sulphur') is usually to combat an excessively alkaline soil rather than as a necessary correction for sulphur deficiency.

The major danger of excessive application of sulphur in any form is that it lowers the pH value of the soil. This is because it always ends up, in a soil which is adequately aerated, as sulphate. In flooded soil sulphur is present in the form of sulphides and, in addition to smelling rather foul, the sulphur accumulates as it deposits out of solution. If such land is drained it may subsequently become so acid, due to sulphate formation, that it will not support plant growth.

Calcium

Particularly involved in the normal development of growing points and in the formation of strong linkages between cells it is only to be expected that calcium deficiency symptoms are concentrated at young growing regions, which often die back and are further manifest by a generally 'soft' growth of the rest of the plant which is very susceptible to rotting. These symptoms are usually seen on strongly acid soils and liming is often used to correct low soil pH values even if calcium deficiency symptoms

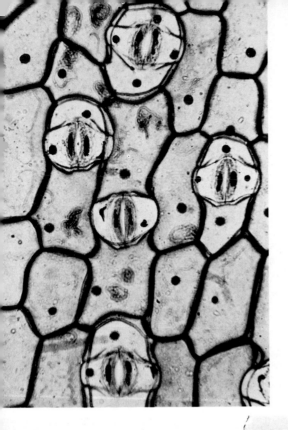

Page 65 (*left*) *Surface* view of the leaf epidermis of *Tradescantia* to show the stomata. Highly magnified; (*below*) section through a leaf of *Erica* showing the outer epidermal layer and darkly stained chloroplast, containing cells underneath and a central vascular bundle—highly magnified.

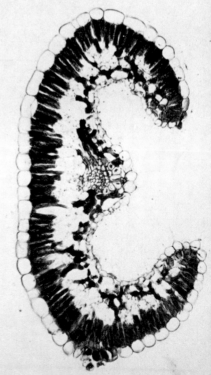

Page 66 (*left*) Bonfire smoke is a potential hazard to health. Although this home-made incinerator allows air to the sides of the fire it does not let any underneath the rubbish and so the fire smokes more than is necessary; (*below*) the distinction between cultivated plants and weeds is not always very clear. This is a cultivated thistle and not a weed.

are not actually present. Its effect in improving the soil structure of clays has already been mentioned. By raising the pH value towards 6, liming also improves the availability of other elements. Very high levels of calcium cause considerable side effects by reducing the availability of elements such as magnesium and potassium. Chalk soil typically needs regular applications of potassium fertilizers while methods of overcoming the iron deficiencies which develop have already been mentioned. Overliming can also cause humus to break down too quickly, damaging soil structure, and lime should obviously not be added to soils which are already neutral or alkaline. Many forms of liming material are available; quicklime and slaked lime act quickly in neutralizing soil acidity while chalk, lime and marl are slower acting. Ground sea-shell, egg-shells and many waste products from industry also contain calcium and can be used when available. Usual recommendations for liming are once every two or three years, with less needed on sandy soil than heavy clay.

Magnesium

This element is part of the green chlorophyll molecules, therefore when it is in short supply a typical symptom is *chlorosis* (yellowing) between the veins, beginning in the older leaves, which also frequently fall off prematurely. This is sometimes seen on acid sandy soils which are also deficient in calcium, and here it is usually cured by adding dolomitic limestone which contains both magnesium and calcium. Another cause of magnesium deficiency is excessive application of potassium fertilizers which causes a nutrient imbalance.

Iron

Although not part of the chlorophyll molecule, iron is associated with its formation, and chlorosis can be caused by iron deficiency. This is often seen on soils with a high lime content where the soil may contain iron but not in an available form. Excessive

E

amounts of the other 'trace' elements in the soil may also cause iron to be unavailable. Iron deficiencies are frequently cured by spraying a dilute (0.1 per cent) solution of an iron salt onto the leaves or putting a chelated ('chemically protected') form into the soil.

Manganese, Molybdenum, Zinc, Copper and Cobalt

Deficiency of these trace metals can also cause characteristic symptoms but often the effects cannot be readily detected except as a reduced yield. The amounts needed by plants are so low that often deficiencies are cured by changing the soil pH slightly rather than actually applying the missing element. Foliar application of dilute solutions is the usual means of supplying the trace elements as soil application would be wasteful, even if it proved effective.

Boron

This element, although not clearly involved in any particular aspect of plant growth, is essential in trace amounts and many 'diseases' are caused by boron deficiency. Death of growing points, breaking down of vascular tissues, heart rots and browning in the centre of fleshy stems are all typical symptoms. A possible explanation may be that calcium utilization is abnormal in the absence of boron. Only very small amounts should be applied to plants or soils otherwise toxicity due to excess may develop.

Sodium

Certain plants benefit from a little sodium (ordinary salt) added to the soil. Many members of the cabbage family such as brussels sprouts, kale and so on benefit, as well as beetroots and celery. No definite role for sodium has been discovered in plants although the element has a close chemical relationship with potassium. Light applications of salt should be given only to those plants known to benefit from it.

Fertilizers

The growth of plants and their subsequent decay form part of the cycle of nutrients occurring in nature. By now it should be clear that organic matter is good for the soil because it contains all the nutrients needed for plant growth in the right proportions, as well as improving soil texture and properties. It is rarely possible to return as much nutrient to the soil in organic form as crops remove. Inorganic fertilizers are usually necessary to replace the total amount. A 'fast-acting' fertilizer contains the elements in a soluble form for immediate use by the plants, while a slow-acting one releases them slowly over a longer period of time for sustained effect. The three major elements nitrogen, phosphorus and potassium are usually added at the same time in the form of an NPK fertilizer. The proportions of the three ingredients are varied to suit particular crops, more nitrogen for leafy vegetables, more potassium for fruits and seeds. If wanted, separate additions of fertilizers containing only one of these elements can be made. If growth is still poor after NPK addition a trace element deficiency may be the cause. A check of soil pH value should be made at this time (if not before) because a modification of this may correct deficiencies without the need to find out which element is actually missing. A chemical analysis of the soil will give information about its nutrient status and will be an economic proposition if a large area of land is being cultivated. Continued use of the same fertilizer year after year can cause nutrient imbalances to develop in a soil and it is clearly wasteful to add nutrients which are already present in adequate amounts. Deficiencies of trace elements are usually indicated by characteristic symptoms, although these may only show on some kinds of plants. Diagnosis, and cure, is often accomplished by foliar spraying of very dilute solutions of the elements, either singly or in combination. When normal growth is restored in a trial plot diagnosis and cure are effected simultaneously and the treatment can be used on a larger scale.

OTHER ENVIRONMENTAL FACTORS AFFECTING GROWTH

Many other environmental factors which affect plant growth could be catalogued and the interactions between them and the primary factors of temperature, light, water and nutrient supply. This present book, however, aims primarily at drawing the gardener's attention to the problems rather than exploring them exhaustively. Some of these other factors will be mentioned in various later chapters. The force of gravity, for example, which influences the appearance of many plants is discussed in the next chapter. Oxygen supply has already been mentioned in connection with respiration and soil structure and will feature again in the chapter dealing with germination. Air humidity obviously affects water loss from shoots and this may determine whether or not cuttings will 'take', so this warrants further attention when we discuss propagation. Carbon dioxide, the raw material for photosynthesis, obviously must be in adequate supply but in common with many environmental factors it can really only be 'controlled' in a closed system such as a greenhouse.

Wind or air circulation normally exerts its effects in rather indirect ways. One obvious direct effect is the mechanical damage caused by high winds. Thin stems will be broken off at weak points; top-heavy blooms obviously need adequate staking; heavy ripening fruits will be torn off. These and the other detrimental effects of winds can be minimized by using sheltered spots for easily damaged plants and providing natural or artificial windbreaks.

Surprising as it may seem to some people a wind-break is better if it allows air to pass through it. If it is a continuous, impervious screen the air will be forced up and create considerable turbulence on the other side. The air should be slowed down by the wind-break, not completely stopped. As is well-known the temperature of a wind and its humidity vary according to the direction from which it comes. Hot dry winds cause desiccation and signs of drought to develop rapidly; cold winds can cause frost damage and carry heat rapidly away from plants so that growth is slowed. Damp winds, on the other hand, are less damaging to plants unless the moisture is heavily laden with salt spray. The effects of 'wind pruning' are mentioned in the next chapter.

POLLUTION

Although we are all familiar with the term 'pollution' it covers a multitude of sins and is difficult to define exactly. The average gardener is unlikely to be troubled by plastics unless he tries to incorporate them into the compost heap. Although they are organic compounds containing carbon they differ from natural materials because they are not easily broken down into their components by micro-organisms, but instead remain unchanged in soils. Burning may reduce them in bulk but often a tarry residue remains which is of no value to the gardener.

Beyond the control of the gardener are the gases and dusts which come over the garden wall such as the sulphur dioxide produced by both domestic and industrial fires. The gas readily dissolves in water and is removed from the atmosphere by rain and transferred to the soil where oxidation produces sulphates. Damage caused to plants will obviously depend on the concentration of sulphur dioxide in the air and how long the plants are exposed to the gas. Low concentrations (even below 1 part per million) cause a gradual breakdown of chlorophyll, resulting in chlorosis and a general suppression of growth. High concentrations cause a rapid chlorosis and then *necrosis* (death) of the cells. Different species of plants differ in their sensitivity: among

71

the conifers larch is particularly sensitive and junipers are very tolerant while broad-leaved trees and shrubs are generally more tolerant than conifers. Gardeners unfortunate enough to live near sources of sulphur dioxide have to restrict their plants to those which are resistant to it. But fortunately hundreds of plants have been tested for sensitivity to sulphur dioxide and the list of plants which can tolerate it is quite long although personal observation can obviously extend it even further.

Certain forms of fluorine are liberated from some industrial refining plants where minerals are heated. The element also occur as a natural component of soils, and certain plants such as members of the camellia family can accumulate very large amounts although these do not appear to serve any useful purpose. The signs of atmospheric pollution by fluoride are again those of chlorosis and necrosis of leaf tissue, particularly at the tips and edges of leaves. Once again different species and even different cultivars can show considerable differences in sensitivity to fluoride.

The term smog was first applied to a combination of fog and smoke in the early 1900s when the toxic component was mostly sulphur dioxide. With the advent of the internal combustion engine and the motor car the nature of smog changed. Sulphur dioxide in the air has been reduced by changes in fuels and improvements in the cleaning up of waste gases from industrial plants. The new type of smog, often known as Los Angeles smog, is completely different because it results from the interaction of sunlight with hydrocarbons and oxides of nitrogen given out in the exhaust fumes of motor cars. The complex products cause plant damage, eye irritation and have many other damaging effects. One of the most toxic components as far as plants are concerned is *peroxyacetyl nitrate* (PAN for short) which was first detected as recently as 1957. Another toxic substance is ozone, which may surprise those brought up on the idea of 'taking the ozone' at seaside resorts for the benefit of their health, but in

towns and cities the concentration is likely to be high enough to
cause considerable damage to sensitive plants, though it needs a
skilled eye to distinguish between smog damage to leaves and
damage caused by other agents. Once again some plants are much
more sensitive than others: spinach, for example, soon succumbs
while cucumber is more resistant. The gardener must choose his
plants to suit the conditions if he lives in an area affected in this
way.

Ethylene is in a unique position with regard to pollution. Not
only is it produced by car exhausts in toxic amounts but it is also
a natural product of many plants and is considered to be a plant
hormone. Among the plant growth processes which are in some
cases regulated by ethylene are fruit ripening and colouration,
leaf and flower abscission and root production. The effect of toxic
concentrations on plants are apparently due to an upsetting of
normal growth patterns rather than to a directly poisonous effect
but huge economic losses have been caused by ethylene pollution. Orchid production in the proximity of heavily polluted
areas like Los Angeles has been stopped because even two parts
of ethylene per thousand million can cause a detectable decrease
in commercial value of the blooms. At only slightly higher concentration 'dry sepal disease' occurs with very premature death
of the flowers. In carnations 'sleepiness' can be caused by ethylene, with the buds failing to open and petals turning yellow and
withering. At one time ordinary cooking gas supplies contained
large enough quantities of ethylene to cause extensive damage to
pot plants in kitchens or in gardens where a slight leak was
occurring from gas pipes. It was even possible to speed the ripening of tomatoes by enclosing them in an air-tight container filled
with coal gas. Natural gas, on the other hand, contains virtually
no ethylene, and it is now possible to buy natural gas heaters for
greenhouses which produce so few toxic gases, even when
alight, that they can be used without a flue.

Other gases, such as chlorine and ammonia, occasionally cause damage to plants but this mostly is the result of accidents to tankers or in factories. Dusts may emanate from industrial plants, such as cement works, or be formed by the burning of vegetation or other waste materials. It is usually difficult to sort out which component of smoke is causing the greatest trouble because incomplete combustion can yield almost every type of atmospheric pollutant, but the dust itself appears to cause damage by blocking the stomatal pores and forming a layer over the leaves which reduces the light reaching the leaf. Heavy rain can wash off the particulate dusts but where oily residues are also deposited, such as on plants growing close to roads, even rain may not be able to wash them off.

The pollution problem caused by pesticides is now a common talking point, even in the absence of reliable facts, and will warrant special attention in Chapter 8.

Chapter Four

Controlling Plant Growth

In the first first three chapters our main attention has been given to the needs of the plant, and the gardener has only entered the picture as an assistant who can help the plant to attain its full development. Now our attention turns more towards the means by which the growth of a plant can be directly affected, to the advantage of the gardener.

GROWTH HABIT

In many plants if the main apex is not damaged it will go on growing to produce a single main stem. The lateral buds will either fail to develop or do so only slowly. This is due to *apical dominance* in which the leader suppresses the development of the laterals. It has been known for a long time that auxin produced in the main apex is transported only down the stem, away from the apex. The simple conclusion was that auxin inhibited the outgrowth of buds and this seemed to be confirmed by the fact that a paste containing IAA would replace an apex which had been cut off and maintain apical dominance. There are many experimental results which do not support this simple theory and

75

current explanations of the role of IAA are rather more complex. If a portion of a stem containing a dormant lateral bud is sectioned and examined under the microscope it is seen that the bud is not connected to the main vascular system of the plant. But when a bud starts to grow it is connected up to the vascular system so that it can be supplied with nutrients and water. Auxin is thought to control this joining up of the vascular system of the stem and bud and to exert its effect upon the nutrient supply to the bud rather than on the bud itself. In the garden the role of nutrition in apical dominance is easily seen. Plants with inadequate nutrition due to low light or poor soil frequently show much stronger apical dominance and develop fewer lateral branches than plants with good nutrition. Nitrogen is particularly important.

Apart from apical dominance the other basic fact which allows control of growth is that buds do not normally develop spontaneously on stems. They occur only in the leaf axils. If the lateral buds are removed as soon as they show the slightest sign of growing a single tall stem is obtained. Alternatively if the apical bud is removed the laterals will be encouraged to grow out and a lower, bushy plant results. There are plants in which this simple rule does not apply. In most conifers, for instance, if the main leader is damaged or removed its place is taken by a nearby lateral branch which changes its direction of growth and grows upwards to become the new leader. Naturally this causes a kink in the system and may entirely destroy the commercial value of the tree.

In woody plants two basic types of branching can be recognized. In monopodial branching the apical bud of any shoot retains its ability to grow over many years, and also its supremacy over the lateral buds. *Forsythia* and *Magnolia* are of this type. The second type of growth is called sympodial and in this the terminal bud either dies off each year or is replaced by a flower and the laterals therefore develop in its place. *Syringa* (lilac) and many flowering shrubs show this type of growth. Most flowers are considered to be modified buds, either apical or lateral and this accounts for their normal positions either at the tip of a shoot —as in *Buddleia*—or in the axils of leaves. Where more than one flower develops from a single bud it is usuallly taken to represent a reduced branch system. Botanists talk about the *transformation* of a vegetative bud into a floral bud. Once this has occurred it is rarely possible to reconvert a floral apex into a vegetative one. Obviously the position of the flowers is going to affect the shape of a plant because certain branches cannot develop as they have been replaced by the flowers.

So far the plant has been discussed only as a system of stems, branches and leaves, but the final appearance of a plant is dependent upon the way in which these are borne with respect to one another. The symmetry of a well-grown plant is related to the way in which the leaf primordia are produced at the stem apex and also to the actual position on the stem from which the mature leaves arise, which usually reflects the situation at the apex.

GEOTROPISM

As soon as a primordium has formed it becomes subject to environmental influences. As an organ develops, its orientation with respect to the axis which bears it becomes apparent. This depends on the relative growth of the upper and lower sides, and applies to both stems and leaves. The rate of growth is determined partly by inherent characteristics and partly by environmental influences among which gravity is of considerable importance. The term *geotropism* is used to describe situations in which parts of a plant assume particular orientations with respect to gravity. Perhaps the simplest response is shown by the main roots of seedlings. These grow downwards, towards the centre of the earth. If we dig up a seedling and lay the roots along the ground, the part just back from the tip will soon show a growth curvature and allow the root tip to grow downwards again. This is an example of positive geotropism—growth towards the gravity stimulus. The opposite of this is negative geotropism, shown by many young seedling shoots which grow away from the ground. The exactly vertical spires of lupin, delphiniums and many other flower spikes are examples of negative geotropism. Let them fall over or be knocked down by the wind and they will quickly respond by a growth curvature which is virtually irreversible. If they are then tied up again in their original position a second response occurs to ensure an exactly vertical stem, but one which now has two kinks in it. Lateral branches, leaves and roots often orientate themselves at some particular angle to the gravity stimulus and are known as plagiogeotropic organs. A special situation of importance to gardeners occurs when the root or stem grows exactly at right angles to gravity, so that a rhizome, for instance, grows along just below the surface.

Geotropism can interact with many other environmental factors. If you have one of the red deadnettles (*Lamium purpureum*) as a weed in your garden, you will be familiar with the conversion from its normal summer form, which is erect and negatively geotropic, to a horizontally growing, creeping form in winter, the change being caused by low temperatures. Even more complex are the responses of that noxious weed ground elder (*Aegopodium podagraria*). The rhizomes maintain a fixed depth in the soil because if they grow into the light they become positively plagiotropic and so grow back into the soil. Their penetration to too great a depth is prevented because high levels of carbon dioxide cause them to turn upwards again.

Many plants have horizontally growing branches which also show a marked flattening—many conifers immediately come to mind. In some of these cases, the most famous of which is *Araucaria excelsa*, the horizontal growth pattern somehow becomes fixed so that cuttings taken from side shoots grow horizontally instead of reverting to the negative geotropism expected of a rooted cutting and growing upwards.

Even familiar flowers are not exempt from the unseen influences of the environment. The botanist considers that regular flowers were present on the earth before irregularly shaped ones. A regular (*actinomorphic*) flower is one which can be cut in half along all the radii from its centre and still give two equal halves. A *zygomorphic* (irregular) flower can only be cut into two equal halves along one line. This type of construction is typical of the pea and orchid families. Many zygomorphic flowers are specially adapted for pollination by particular insects and while the shape of many zygomorphic flowers is unaffected by gravity some, including *Freesia* and *Digitalis* (foxglove) change from a zygomorphic to a regular form if grown in such a way that gravity is acting equally on all sides. The botanist usually uses a *klinostat* for this purpose, on which the plant grows horizontally and is slowly rotated. Peloric (regular) forms of *Linaria* (toadflax) and *Digitalis* flowers have been known for a long time and in these the regular form occurs even under normal growth conditions. As a completely separate response almost all zygomorphic flowers assume a definite position with respect to gravity—obviously an essential feature if a bee or other insect is to land on a specific part of the flower to effect pollination—the whole complex shape of a highly evolved flower would be useless if it was borne upside down.

CULTURAL PRACTICES

So far nothing has been said about the control of plant growth, which is the main point of the chapter. The reason for this is that one cannot sensibly control what one does not understand, but we are now in a position to consider such simple gardening practices as pinching out and disbudding.

Suppose we have a seedling bedding plant with a single main stem and when planted out we require a bushy plant with plenty of lateral branches bearing flowers. These laterals can only come from buds in the axils of the leaves (see below). It is therefore a mistake to pinch out the main growing apex before a reasonable number of leaves have been formed. On the other hand, it is not correct to let the plant become drawn in the normally rather dim light of the greenhouse so that the internodes elongate since once this has happened they cannot be shortened. As mentioned previously insufficient nutrients in the soil, a poor water supply or restricted photosynthesis will hamper lateral bud development. So we usually plant out and allow a little time for establishment before pinching out the apex. Where several stems already exist the pinching out of the apex of each will stimulate further branching.

Disbudding can be considered as the opposite to pinching out. If we want to produce a single tall stem rather than a bushy plant

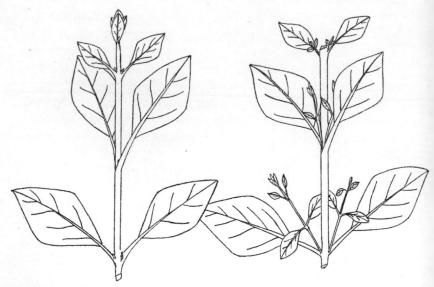

The effect of pinching out, which liberates the laterals from apical dominance

we can remove all the lateral buds as soon as they are large enough to see. Once the original bud has been completely removed a new bud will rarely form in its place. Apart from giving a plant a convenient shape, the main reason for disbudding is to concentrate the resources of the plant into the part which we are going to harvest such as the fruit in the case of tomatoes. Remember that the organic substances of the plant come from the products of photosynthesis in the leaves and if all the buds are removed and, perhaps due to drought, all the leaves drop off, the plant will be in serious nutritional trouble because new leaves can only come from the primordia in buds.

Flower growers have long practised disbudding in order to obtain larger blooms. With dahlias, for instance, the lateral flowers are commonly removed so that only a single bloom develops on a stalk. This will then be larger than if all the blooms were allowed to develop. Occasionally the reverse is done, removing the central, largest, flower bud to obtain a 'spray' for flower arranging. Fruit responds in a similar way to thinning out although here nature often also takes a hand by having a natural thinning when some of the fruit drops off.

The responses to disbudding are often described as *compensatory correlations*. If something is lost, what remains gets larger. Conversely if something remains it prevents the development of new structures. For example, in plants in which flowering and fruiting occur throughout the season, when a certain number of fruits have set no new flowers will develop until the growing fruits are removed. The number of fruits which a plant can bear obviously depends on its nutritional status. In the case of the tomato mentioned previously vegetative growth stops after a large number of fruits have formed and if the flowers and fruits are continually removed a much larger vegetative plant is produced.

This same principle is used by gardeners who are careful to remove dead flowers and young fruits from plants grown for flowers alone. By this means further and more prolonged flowering is stimulated. If the task is neglected flowering soon stops as the plant directs its energy into fruit and seed production. The influence of one part on another is even shown by leaves. If a leaf is removed from a suitable plant and rooted as a cutting, it will frequently grow to a greater size and live longer than if left attached to the parent plant.

CHEMICAL TREATMENTS

Many of the examples quoted could be taken as evidence that it is the nutritional aspect alone which controls growth. This, however, would be an incorrect assumption. Much evidence is available that specific hormones influence correlation. The role of auxin in apical dominance has already been discussed, but what of the other plant hormones? Are they of any use in controlling plant growth? It has been clearly shown that the application of cytokinins to lateral buds can stimulate them into growth, releasing them from 'apical dominance' caused by the main apex itself or auxin which is replacing it in a botanist's experiment.

The lateral buds of *Coleus, Helianthus annus* (sunflower), *Pisum sativum* (pea) and others have all been stimulated into growth by direct application of cytokinins to them. Unfortunately this is not the complete story because the lateral shoots only grow about an inch or so and then stop. The application of IAA (auxin) to the apex of these lateral shoots can make them grow out to a normal length but here again there are problems because the leaves still remain small and do not expand properly. Clearly a spray of these hormones cannot yet substitute for pinching out!

At one time it was thought that gibberellins would be of great use in horticulture because they stimulate the elongation of shoots. This was of little value however because the increase in length was usually at the expense of a reduction elsewhere, as we

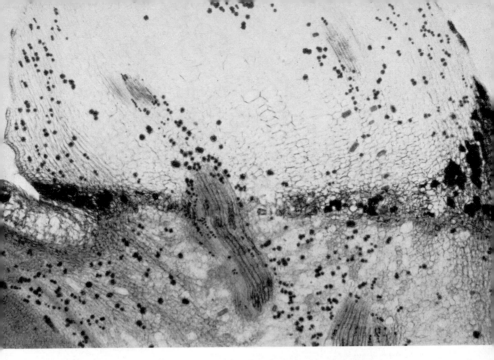

Page 83 (above) This section of a leaf base of Virginia creeper shows the protective abscission layer forming just before leaf-fall—highly magnified; *(below)* this rose is badly in need of pruning: branches criss-cross in all directions and dead flowers remain as reservoir of disease for next season.

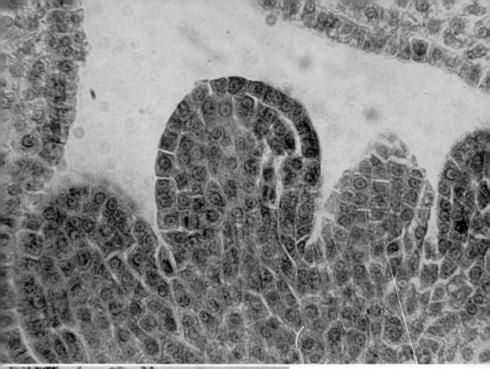

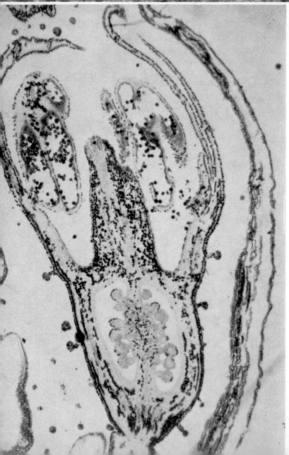

Page 84 (above) Section of
a young *Berberis* flower bud
showing some of the
primordia which will grow
into parts of the flower—
highly magnified; (*left*)
section through a flower of
Ribes showing the many
ovules in the ovary, part of
the style rising above it and
the anthers containing
pollen—highly magnified.

would expect from our knowledge of correlation. When examined in connection with apical dominance gibberellins usually increase main shoot growth and keep the lateral buds in the dormant state. This is probably a reflection of the general principle that the more actively a shoot is growing the greater will be its power to dominate the growth of other shoots. The suckers thrown up from rose rootstocks provide a familiar example with their vigorous growth, sapping the strength of the budded top growth.

Recently a group of synthetic chemicals have been developed which are called morphactins. These do not appear to be naturally occurring substances like the plant hormones but neither are they as toxic as the synthetic weedkillers. The morphactins can produce a wide range of effects, depending on the way in which they are used, and have the advantage of being converted rapidly into inactive substances in the plants. For the moment it is sufficient to say that morphactins can abolish apical dominance and bring about the production of compact bushy plants having a dwarf appearance. One disadvantage of the morphactins is the variety of effects which are also produced but other synthetic substances are known which reduce growth more specifically. These are grouped together as growth retardants although this does not imply any chemical relationship between them. As usual they are known by initials or cryptic names because the full chemical name is far too long and complex. The growth retardants have already found commercial applications. Familiar names to those in the business are CCC, B9, phosphon and maleic hydrazide, each with its own particular characteristics. The preferred method of application to prevent damage is usually by the use of modern, specialized, ultra-low volume fogging machines. These substances reduce the growth of the treated plants, which are usually pot flowers such as asters, chrysanthemums or marigolds. The shape of the plant is little effected except that it is dwarfed, giving it a commercial advantage over tall straggly pot

85

F

plants. The effect is not permanent and offspring from the plants will not be dwarf. It is an effect on the growth of the treated plant alone and is not inherited in the way in which a true dwarf cultivar is.

PRUNING

Pinching out and disbudding are terms usually applied to plants with soft growth where the fingers can be used. When mechanical aids such as secateurs and saws have to be used, because the plant is woody, we are in the field of pruning, topiary and the like. The processes we are attempting to control are the same as those already described.

In the young shrub we want to develop a good, balanced shape; as it becomes older the aim will be to stimulate flowers or fruit; finally we will desire to keep its size within bounds. For hedges the aim will normally be a dense growth which extends down to the ground. Perhaps the most important question before pruning is to decide whether the plant flowers and fruits on the current season's growth ('new wood') or on last year's growth ('old wood'). As a general rule plants which flower early in the year, like *Forsythia*, must be doing so on old wood, while those flowering in the autumn, such as *Buddleia* are doing so on the new wood. Of course there are plants like gooseberries and some clematis, which flower on both types of wood. Briefly the principle of pruning is normally to do it when one cycle of growth has finished and before the next starts. Thus as soon as raspberries have fruited the old canes are cut down to the ground to obtain maximum growth of the new shoots which will provide next year's fruit. Similarly most early flowering shrubs are pruned immediately after flowering to obtain maximum growth for next year.

The late flowering plants are pruned during the dormant season, often just before regrowth starts. Those plants flowering on both old and new woods can be manipulated by pruning to give early or later flowering. Plants like the roses which produce plenty of new wood can even be pruned by stages as the season progresses in addition to the major pruning at the beginning, during or at the end of the dormant season.

The actual techniques of pruning differ, depending on the species, but there are a few common features. Only sharp tools should be used because rough, jagged cuts provide a resting place for disease spores and the water which encourages their growth, as well as leaving the plant with a greater area to heal. Large cuts must, of course, be treated with a wound-sealing compound to reduce the risk of infection. Cuts should always be made just above a bud or level with the joint onto a larger branch because small stubs invite infection and normally die back anyway. Obviously all dead or feeble growth must be cut out and any woody pieces must be burnt as soon as they are dry. If they are left around they will act as bases for infection of the tree, and woody material will not decay in the compost heap. The shape of a shrub, tree or wall climber should be planned and shaping begun as early as possible. If a plant has been neglected it will naturally take a little time and probably great forethought before it can be coaxed into an attractive shape. (See bottom photograph on page 83.)

Shears are only suitable for trimming hedges and one or two shrubs. With shears one cannot control the exact position on the shoot which is being cut to ensure that it is always just above a bud. Therefore only those plants which respond by sending out laterals readily and which do not die back from the cut ends will make a good hedge for close trimming. The high development of the art of topiary bears witness to the ability of some plants to respond well to this treatment. Because of the human decisions needed in pruning no chemical is really conceivable which can do the job for us but once a hedge has been shaped, however, it is possible to use growth retardants to limit the amount of clipping which is needed. The decision here is usually an economic one but great care is needed in the use of growth retardants if permanent damage to the plants is to be avoided.

Mowing grass is, in a way, another example of removing the apical growth to stimulate lateral growth and the good lawn grasses are usually those which spread out to form a close mat. Once again growth retardants can, with care, be used to reduce the labour of control.

A certain amount of natural pruning takes place if plants are completely neglected. The tangled branches at the centre tend to be weak and die out, while fresh growth occurs round the periphery of the plant where light and air are available. The lower branches on many trees often die off, particularly if the trees are crowded together so that competition for light, and presumably also for nutrients, is intense. The 'typical' shape of a tree is only attained where the plant is growing with little competition from other trees. Wind-pruning is a special case of natural shaping in which the prevailing wind, because of its temperature, dryness or high content of salt inhibits or even actually kills the young growing buds on the windward side. The result is the typical wind-swept shape to be seen along any coast or exposed position where the wind is predominantly in one direction.

OVERCROWDING

Mention has just been made of competition and this is a problem with which every gardener is familiar. No plant can reach its full potential if it is grown in overcrowded conditions where competition is bound to occur for water and nutrients in the soil and one plant is bound to suffer shading from another. The longer competition is allowed to continue the greater will be the damage to the form and growth of a plant. When seeds are planted they will germinate even if they are so close that they actually touch but the young seedlings soon begin to exert an influence on one another, and since the seed normally has a supply of reserves the major competition will be for light. Shading of one by another rapidly leads to partial etiolation with long, weak internodes. So we have the familiar tasks of thinning out and planting out. Young seedlings are quite resistant to damage caused by moving and the sooner they can be given a reasonable space the better. They do not like to be too far apart however, presumably because there is then no mutual protection from drying winds or the baking effects of the sun on the soil. If the plants are left too close, on the other hand, one can only expect twisted growth, stunting of those seedlings which do not get away quickly and generally poor results.

THE SIZE OF PLANTS

Moving on from ways of controlling growth, the most sensible way of obtaining a plant of the size and shape which we want is to select a cultivar (variety) which has the correct characteristics. A good nursery catalogue will give all the information needed to plan the plants of a garden for maximum visual impact and convenience of cultivation. Few people will have failed to notice that certain cultivars of rose are weak growers while others, like 'Queen Elizabeth' are rampant. There are bushes, climbers, ramblers, miniatures and so on. A garden planned without thought for height or vigour is probably even worse than one without thought for colour clashes. These characters now under discussion are hereditary ones, passed on as essential features of the cultivar. Obviously a plant cannot reach its full potential unless grown under correct conditions, but the plant catalogues naturally assume that we do this.

Much scientific work has been carried out on dwarf forms of some crop plants. With these it has been found that in the tall forms there is the usual apical meristem producing new cells, and the zone of elongation where cells (and hence stems) grow longer. Between these there is often a *sub-apical meristem* which produces large numbers of cells for the stem. The length of the stem is dependent upon the activity of this sub-apical meristem. If it is active we have a tall plant and when it is inactive a dwarf plant is the result. The gibberellins are very much concerned with the activity of sub-apical meristems and if they are, for any reason, not present in sufficient quantities the meristem will be inactive and a dwarf plant is produced. It follows that some (but not all) dwarf plants can be made to grow into tall plants by applying gibberellins to them. Most of the growth retardants seem to work by preventing the normal functions of the plant's own gibberellins, even though they are chemically completely unrelated. In the opposite case the effect of the growth retardants can often be overcome by applying gibberellins to the retarded plants. The conversion of a dwarf plant into a tall one again does not change the hereditary characters of the dwarf plant—all the offspring will be dwarf unless treated with gibberellins. It must be emphasized that not all dwarf plants can be converted into tall forms by gibberellins—there are many different reasons why a plant may be dwarf.

Although few gardeners seem to actually use them most have heard of the dwarf fruit trees, particularly apples, which can be obtained by grafting onto a 'dwarfing' rootstock. Almost all commercial tree fruits are obtained by grafting onto a suitable rootstock which will give the size of tree desirable for the particular orchard. The way in which a rootstock can effect the shoot system has been subject to much experimentation. The main conclusions appear to centre on the idea that either the water supply or nutrition of the shoot is reduced. There are usually fewer water-conducting vessels in dwarf rootstocks, and also the transport in the phloem between rootstock and scion of organic substances is restricted. Actually a dwarfing effect can also be produced by taking off a complete ring of bark and turning it upside down before replacing it. This also disrupts the passage of nut-rients in the phloem. The mere fact of using an inherently smaller-growing rootstock will naturally reduce the growth of the above-ground parts.

The ultimate in dwarf trees must surely be the bonsai—a term which implies growth in a pot. The essence of bonsai culture is to obtain a tree which resembles a naturally grown one except for its greatly reduced size. The principles are simply to restrict root and shoot growth, by careful root and shoot pruning, whilst still maintaining the relative proportions of the parts of the tree. Although any tree can be dwarfed by suitable culture some are naturally more suitable than others for this type of manipulation.

ABNORMAL GROWTH

Some attention must now be paid to abnormal growth. We have all seen this at some time and probably speculated on its nature and origin. One of the most common types for the gardener is that which results from incomplete fertilization, with a conse-quent development of lopsided fruits. This is explained in the chapter on flowering and fruiting. Any kind of damage to the plant frequently results in the development of abnormal growths, particularly if it occurs during an active growing season. The ab-normality may be as simple as a return to juvenile-type foliage but perhaps the most striking is the development of fasciation, which is a flattened or even a ring-shaped outgrowth, usually bearing dozens of leaves or flowers. Two explanations have been suggested, firstly the growing region at the apex may expand sideways and secondly a number of primordial regions may fuse together to give the larger growing point. Botanists tend to restrict the term fasciation to the first type of change but, since the type cannot be determined simply by examining the fasciated structure there seems no point in the gardener drawing such dis-tinctions. Heavy pruning of *Salix* (willow) and many other trees and insect damage to the apex can cause fasciation. Certain plants have an inherited tendency to produce fasciated stems. The flowering stem of the hyacinth is said to be of this type. *Celosia cristata* (cockscomb), *Phlox drummondii* and other plants may also show this tendency, but to a lesser extent.

Treatment with weedkillers and other chemicals often results in distorted growth, fusion of parts or the development of organs, such as roots, in regions where they would not naturally occur. The recently studied morphactins are particularly effective in causing abnormal growths. Using these the leaf primordia may be made to fuse into cornet-shaped funnels and groups of individual flowers joined into one fused 'flower'. Changes in the individual number or appearance of the parts of flowers frequently occur naturally so it is not surprising that such changes also result from chemical treatment. Abnormalities of growth caused by damage are the result of an upset in the development of the individual plant and fortunately, or unfortunately as the case may be, cannot be passed on to the offspring, but the tendency to produce particular types of abnormality spontaneously as in the peloric flowers of *Linaria* and in the examples of fasciation quoted above can be inherited.

Chapter Five

Seeds and Seedlings

The production of seeds is one way in which a species of plant may be perpetuated or its numbers increased. Essentially a seed consists of an embryo protected by one or more enveloping layers. The embryo is a miniature plant with a root and a stem apex plus one or two cotyledons or in conifers sometimes more. The only other requirement is a reserve of food to support the early growth of the seedlings. There are two places where the food may be stored—inside the embryo, especially in the cotyledons, or outside as a separate 'endosperm'. The food reserve itself is usually mainly carbohydrate or fat, although in a few seeds proteins may form a large part of it. There are also present the various mineral elements and other organic substances needed to give normal growth until the seedling can become self-supporting.

THE FORMATION OF SEEDS

Although seeds can arise in other ways, a typical seed is the product of sexual reproduction. Pollen grains fall onto the special receptive stigma which is associated with the ovary. There they germinate and a pollen tube grows into the ovule. The nucleus

93

of the egg cell and one of the nuclei in the pollen tube fuse together during fertilization. (See photographs on page 117.)

If fertilization does not occur the ovule is usually (but not always) incapable of developing into a seed. After fertilization the egg soon starts to divide into separate cells, the divisions following a very ordered pattern, the exact details of which depend upon the species of plant. Gradually the plant embryo takes shape. This cannot be seen because it takes place deep inside the ovary just as a human baby develops in the womb. Having made that comparison it should be pointed out that what is called the 'ovary' of a plant is equivalent to nearly the whole of the female animal's reproductive system and not just the animal's ovary—just another example of the confusion that can and does occur if terms are not properly understood. (See photographs on page 118.)

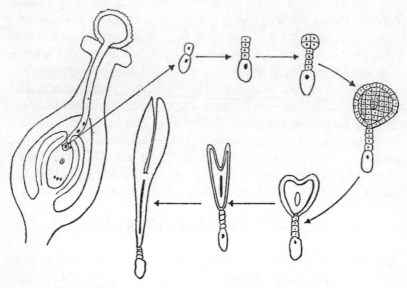

The development of a plant embryo from fertilization to maturity

Seeds and Seedlings

Botanists have been able to remove developing embryos from plants and grow them on sterile culture media. By this means they have found some of the requirements for embryo growth. A fully developed embryo can be grown in the light on a simple nutrient medium which contains only inorganic salts and it is therefore self-supporting like a mature plant. If the embryos are taken at a younger stage, before they mature, the nutrient medium has to be supplemented with other substances: various concentrations of sugar may be needed, plus vitamins, amino acids and plant hormones. Coconut milk is often used as a cheap additive because it contains many of the essential nutrients. Incidentally, coconut milk is a liquid endosperm and would naturally be used by the germinating coconut seed to support its growth. One interesting result of this type of study is the discovery that some non-viable seeds which are produced when plants are crossed have normal embryos but the endosperm does not develop properly and the embryo subsequently dies. Using embryo culture it has been possible to obtain hybrid plants from normally infertile crosses. This only applies to some infertile crosses, not to all of them, and in any case the techniques for embryo culture needs considerable skill. Most orchid growers make use of a technique similar to that used in simple embryo culture when they grow up seedlings in flasks. The embryo of orchids is very immature when the seed is shed and under natural conditions usually needs the help of a special fungus which supplies it with certain essential nutrients before it will grow.

To the botanist the term seed has a restricted meaning; what the gardener calls a 'seed' often has other protective layers either around or even fused firmly to the true seed coat. In this book, however, 'seed' is used in the meaning of the gardener rather than that of the pedantic botanist.

When examined in detail with a hand lens or low-powered microscope many seeds are seen to be of exquisite beauty in the patterning of ridges and depressions on the surface. From the gardening viewpoint the size of the seed is of more importance than these surface markings which are of little relevance. Incidentally, pollen grains often show surface patterns, visible under the microscope, which also appear to have little biological value except to help man in identifying the pollen! The size of a seed puts a limit on the amount of food reserves it can contain. It follows logically from this that the smaller the seed the more quickly the seedling becomes self-supporting. This means that while the seedlings from large seeds have enough reserves to fend for themselves for some time and overcome early competition from weeds, the seedlings from tiny seeds need special care and protection to get them successfully over the critical first days of life.

GERMINATION

Before germination can occur a seed must take up water. The water content of a dry seed is quite low, often somewhere between 5 and 10 per cent and germination will only occur after this has risen to between 30 and 50 per cent or even higher. Obviously the uptake of water can only occur if the seed coat is permeable to water or has at least one spot in it which allows the passage of water. The fact that a seed swells is not proof that it is alive because even dead seeds will take up water. This is because the uptake of water is not due to any special living reactions but simply to the absorption of water by the dry substances in the seed—rather like a piece of dry gelatine swelling when it is placed in water. The initial swelling of seeds is due to the process known to botanists as imbibition. Quite enormous 'suction' forces can be demonstrated during imbibition and a seed can extract water from what may appear to be a relatively dry soil. Its subsequent germination will need a more plentiful supply of water as the growth of roots and shoots is a biological process which does not possess quite the water-absorbing power of the dry seed. Assuming the seed is ready to germinate, the process starts very quickly. The respiration rate rises even before any visible indication of germination can be seen, and this necessitates an increase in the amount of oxygen taken up by the seed. One way of damaging a seed it to let it soak for too long under water because oxygen is not very soluble in water and germinating seeds can use up all that is present, at a faster rate than it can be replaced by dissolution from the air. Many seeds will not germinate under water but large seeds probably benefit from a few hours soaking as they will then be already swollen before planting, which may give them a slight advantage. The water in which seeds have been soaked is best discarded as it will contain organic substances dissolved out from the seeds and seed coats, and these encourage micro-organisms to grow, placing the germinating seed at risk; they may even contain substances which, in their own right, can inhibit the process of germination.

Apart from swelling, the first visible sign of germination is normally the putting out of a root as very few seeds produce a shoot first. This first root is fully equipped with the gravity sensing system which has already been mentioned in the previous chapter. It therefore makes no difference which way up a seed is planted as the root itself will curve down into the ground. In seeds like the broad bean it is possible to see which way the primary root is pointing before we plant the seed but there does not appear to be any disadvantage in making the root curve through 180 degrees before it can grow in its preferred direction. Most roots are unaffected by light but a few, like those of white mustard, are clearly negatively phototropic and grow away from light. Most roots, however, will grow towards moisture and this tendency is usually sufficient to overcome geotropism if the continued downward growth of the root would cause it to enter a region with less available water.

The direction of growth of the shoot as it emerges from the

seed is also influenced by gravity, but the response is a negative one so they tend to grow up, out of the soil. Most shoots do not appear to respond to moisture and only rare examples to the contrary, such as flax and potato shoots which grow away from moisture, have been recorded. With shoots light can overcome the response to gravity, but to do so it must also be brighter on one side when it then causes the phototropic response which has been discussed in a previous chapter. When a shoot first emerges from the seed it is usually still under the soil and therefore in the dark and will therefore be subjected to conditions causing etiolation and show the type of growth typical of a plant in the dark. This means a rapid and extensive elongation of the inter-nodes and a failure of the leaves to expand. All these features assist it in rapidly reaching the light so that it may play its role in the growth of the plant. In some plants, such as peas and beans, there is a further feature of etiolation which is of advant-age, namely the formation of a 'plumular hook'. The *plumule* is the embryonic stem contained in a seed and the 'hook' implies that the tip is bent over so that the side of the stem and not the delicate apex is pushed through the soil. When exposed to light the plumular hook soon straightens out to give an erect seedling.

It has been known for a long time that if the embryo is re-moved from a seed the endosperm does not break down of its own accord, even when soaked in water. When an endospermic seedling is examined it is also obvious that some part of it is pressed close to the endosperm.

One of the earliest detectable changes in any germinating seed is an increase in the activity of its enzymes or the production of new ones which are not detectable in the dry seed. Once the reserves of an endosperm have been broken down into simpler substances these can be absorbed into the embryo. It has recently been discovered that some embryos liberate gibberellins which then cause cells in the endosperm part of the seed to produce the enzymes which break down the food reserves.

VIABILITY AND LIFE SPAN

Returning now to the gardener's handful of dry seeds, is there any way of telling whether they are alive or dead? This question may seem too simple even to ask but is, in fact, one of the most difficult to answer. Clearly we can plant them and if seedlings come up the seeds must have been alive. But what if they do not come up? Certainly they may be dead but often they are still alive and merely dormant because they have not been subjected to some special treatment which they need before they will germinate.

The packets of vegetable seeds which the gardener buys are marked with words that indicate compliance with the Seeds Act, 1920. This means that they were in good condition with regard to percentage germination and freedom from weed seeds, when they were put into the packet. The date of packeting is also given on the packet. What happens after than can influence the percentage germination so it is a sensible precaution to buy the seeds from a reliable shop which treats them with a certain amount of respect.

Because of the problem of dormancy it may not be possible to carry out a meaningful germination test with some seeds. Botanists have therefore searched for other means of testing for viability. A common reagent used is 1 per cent tetrazolium salt, which is turned red by the embryo if it is viable. There is not always a 100 per cent correspondence between this test and an actual germination test, even for seeds which do not show dormancy. One point which should be made is that not all seeds contain an embryo, even though they may appear normal from the outside. Certain families are notorious for this, particularly plants of the parsley family. Parsley, carrot and other seeds from this family often contain 20-30 per cent of such seed when collected.

Even under the best storage conditions the viability of seeds will decrease with time. Among the most important outside factors controlling the loss of viability are the temperature of storage and the moisture content of the seeds. With a restricted type of seed, such as cereals, it is even possible to predict how long the seeds will remain viable from a mathematical equation. Generally seeds must be dry in order to keep well but there are exceptions. The seeds of *Acer saccharinum* (river maple), for example, rapidly lose their viability when dried. Regarding temperature, the higher it is the more rapidly will seeds lose their viability but dry seeds are much more resistant to high temperatures than growing seedlings or plants and there is generally nothing to be gained by storing seeds in a refrigerator rather than a cool cupboard. On the other hand there is no sense in deliberately exposing them to high temperatures above a heater or in the sun.

The life-span of a seed depends very much upon the species of plant. As might be expected a seed which has been bred for a rapid high germination rate does not usually live for very long if not planted. The seed of wild plants, which includes weeds, can often live in the soil for many years. Some botanists have prepared lists of 'records' for longevity and hard-coated seeds are usually amongst the leaders in this table with many examples of a life over 50 or 100 years. The extreme case which is quoted is that of the Indian lotus, the most cautious estimate being put at between 200 and 400 years with a possible maximum of 1,000 years if the results of radiocarbon dating methods are to be believed. There is no reliable scientific evidence to support the myth that 'mummy' grains from the ancient Egyptian tombs were still viable when the tombs were opened. At the other extreme are many tropical seeds, such as those of the rubber plant (*Hevea*) and tea (*Thea*) which live for less than a year. Sealing into airtight containers can often improve the longevity of seeds—indeed special packets are marketed by at least one seed firm to give this type of protection. Seeds of grapefruit and orange which are allowed to dry out at room temperature soon lose their viability, but if kept in the outer compartment of a refrigerator at 5°C (41°F) they will usually still germinate even after a year. The exact details of optimum storage conditions for particular seeds are sometimes difficult to discover because the information is scattered in various scentific books and periodicals. In addition, most experiments are carried out on agricultural seeds, weed seeds or the seeds of wild plants because the seeds of garden plants have rarely been available for long storage experiments. In the absence of exact information the gardener must use reasoned judgement while storing seeds.

DORMANCY

The term dormancy can be applied to any organ or stage of development of a plant in which active growth is temporarily halted. Many different definitions of dormancy exist as well as other terms which mean more or less the same thing. Many seeds are dormant when they are first shed from the parent plant and will not germinate until they have been kept for a certain time. Such seeds are said to require 'after-ripening' and the storage time may need to be anything from a few weeks to several months. After-ripening can occur even in dry storage and botanists do not know the nature of the changes which take place during it although this is partly a result of the fact that when an explanation is found for the dormancy of a particular seed it is immediately placed into some other category of dormancy! With certain seeds it is possible to speed up after-ripening by storing the seeds at a high temperature. The seeds of some species of *Malva* for example become non-dormant after two hours at 70°C but these same temperatures may kill other seeds. With some seeds the dormancy is only relative so that the fresh seeds will germinate if we can find the right conditions. Fresh barley, for example, will germinate at 10°C but not at 15°C. When

101

G

after-ripening is complete the seeds will germinate over a much wider range of temperatures. Usually dormancy due to a requirement for after-ripening can be overcome by removal of the seed coats before planting.

Many seeds from temperate plants require a period of low temperature before they will germinate. This chilling requirement is usually related to winter conditions and is one way in which autumn germination is prevented, allowing the more resistant seed to over-winter rather than the delicate young seedling. Fresh seeds of apple, rose, peach and many others will not germinate if planted directly at 20°C. They will germinate, however, if the moist seed has been previously kept for several weeks at a low temperature, normally somewhere between 0°C and 6°C. Actually freezing is not required and dry seed does not respond to the treatment. The term *stratification* is sometimes used to describe the low temperature treatment of seeds and was originally derived from the placing of layers of seeds in shallow pits in the soil to be subjected naturally to the low temperatures of winter. Another term with a related meaning is *vernalization*, first used by the Russian botanist Lysenko. Although sometimes used indiscriminately the term is best restricted to those cases where low temperature promotes flowering.

In a few cases it is possible to make a seed which normally requires chilling germinate under warm conditions. The seedling, however, remains stunted until it receives a cold treatment (see opposite). In greenhouse experiments where the temperature was kept high the resulting 'dwarf' has lasted as long as ten years without turning into a normal plant. In individual species further complications can occur. Acorns germinate in the autumn by putting out a root and then they stop and will not grow any more until the young shoot has been chilled. The two-year seeds of *Convallaria* (lily-of-the-valley) need a first chilling to produce the root and then a second one to overcome epicotyl dormancy.

One of the changes which has been found to occur when seeds

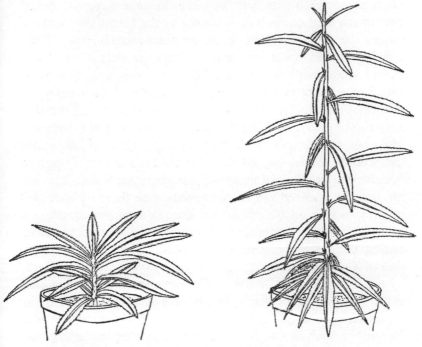

A peach seedling that has not undergone a period of low
temperature remains dwarf until it is chilled

are chilled is a rise in the amount of gibberellins present. It is
therefore not surprising that application of gibberellins can cause
germination in the absence of low temperature treatment. Several
other substances can also break this type of dormancy, including
such simple things as potassium nitrate and complex ones like
kinetin (which is a cytokinin).

Perhaps the type of dormancy most obvious to the average
gardener is that caused by an impermeable or 'hard' seed coat.
These occur in certain members of many different families of
plants. Firstly the coat may restrict the entry of water to the
seed. Such coats are very hard, resisting surface damage, and are

often also covered with waxy material. In some there is a special part of the seed coat which is known by the formidable name of *strophiolar cleft* and when the rather looser material which plugs this is removed water can enter through the cleft. Other hard-coated seeds may have no special cleft region and only allow water to enter after the coat has been damaged by abrasion or microbial attack. The diffusion of gases into and out of the seed may also be restricted. This probably leads to a rise in the amount of carbon dioxide and a reduction in the amount of oxygen inside the seed. It cannot be said that this has been absolutely proved because the problem of analysing the gas inside a seed coat has not been fully solved yet. In many seeds, even those without hard coats, storage in an atmosphere enriched with carbon dioxide can cause or prolong dormancy. Alternatively in some other cases raising the amount of oxygen will break dormancy. Dormancy due to hard seed coats can be overcome, depending on the seed, by simply shaking it; by scratching with a knife; by cutting off a small part near the embryo; by splitting the seed coat or by its complete removal. Botanists have found other ways which may be quicker, such as washing in alcohol or soaking in concentrated sulphuric acid but they do need more care and attention to details during the treatment otherwise the embryos can be killed.

Certain seeds are sensitive to light. The germination of some is promoted by light, for example lettuce and tobacco, but in others such as *Nigella damascena* and *Helleborus niger* germination is inhibited by light. Many others are completely unaffected by light. Even those which do respond show very complex inter-actions with other environmental factors, particularly temperature which may totally alter the response to light. The amount of light needed is extremely small and one-tenth of a second of bright light may be enough for some. The pigment involved in light perception is called phytochrome. Natural light contains both red and far-red components and the seeds which germinate in light are especially sensitive to red light while those which are inhibited by light are more sensitive to the far-red component. Once again certain chemicals such as the gibberellins and nitrates can replace the need for light and removal of the seed coat occasionally eliminates the need for light, as in the case of birch seeds.

With some seeds it is possible to induce 'secondary dormancy'. This could be considered as the opposite to after-ripening. Various factors can induce this, for instance too high a temperature with perhaps some other requirement not at the correct level, or it may happen to seeds which need light but are kept in the dark in the soaked state. This secondary dormancy may require a period of chilling to overcome it. Some botanists consider the state to be identical to 'natural' dormancy.

The remaining problem with regard to dormancy is the presence of germination inhibitors. Many seeds contain substances which inhibit germination and many substances not actually found in seeds can also inhibit. The mere presence of a high concentration of a substance, such as salt after sea flooding or sugars from surrounding fruit pulp, may be sufficient to prevent germination simply by its osmotic effects. Other natural substances such as coumarin and various phenolic compounds are often present in seed coats and can inhibit germination at low concentrations. Many seeds from the rose family contain cyanide-releasing substances; from the cabbage family come examples of ammonia and mustard-oil liberation, and so the list grows longer. Many of these substances also occur in fruits, leaves, stems and roots so it is not surprising that a variety of plant remains in the soil can act as inhibitors if they are not properly rotted—a good reason for composting any vegetable matter before it is incorporated into a seedbed. It is now known that another of the natural inhibitors of germination is abscisic acid which was mentioned previously in connection with leaf and fruit abscission (page 31). Many weed killers, such as 24-D, will inhibit the germination of seeds, both of the crops and the weeds.

In fact there are plenty of reasons why a seed may not germinate even though it is alive. Often more than one type of dormancy will occur in a single seed. How is it possible then to get any seed to germinate? Fortunately most seeds do not give too much trouble. If the seeds are bought in a packet the minimum of trouble can be expected. All vegetable seed must be tested and meet high standards and a reputable firm will also test those seeds which are not actually covered by the regulations. Any special requirements will usually be noted on the seed packets. When we try to grow those plants which are not normally sold as seeds trouble can be expected. It may be that it takes too long to obtain a reasonable plant from seed for there to be any commercial interest, or alternatively that germination is just too erratic and uncertain to risk selling the seed. Suppose you have some seeds which you really want to germinate, you should first try and obtain information about germination conditions for them or for some closely related plant. If no details are available then one just has to use one's judgement. A seed coat that looks too thick or waxy to let water through must be treated in some way to ensure that air and water can get in to the embryo. Does the plant come from a tropical area where a high temperature for germination would be reasonable, or is it from a temperate or alpine plant? With seeds that have not germinated within a month or so, begin to suspect that a chilling treatment of six or eight weeks may be needed. This can be provided in the ordinary domestic refrigerator—other users being willing! When all this fails it could mean that there is some innate dormancy or that the seed is dead, but dead seed will probably have been attacked by micro-organisms and decayed away by the time you look for it in the soil. If the seed still looks as good as new it may be time to try gibberellins or anything else which is to hand!

CONDITIONS FOR GERMINATION

The four essentials for good germination and seedling growth are: (a) an adequate water supply, (b) the correct temperature, (c) adequate aeration and (d) an adequate light supply. How can these be provided? The supply of water is intimately bound up with the structure of the soil. Since the seed composts which can be bought have been carefully worked out to provide the right conditions it is only common sense to buy them or to make up your own to the same specifications. The essential requirements are a fine crumb structure which does not break down with watering and which can hold water well so that it does not dry out too quickly. In the traditional John Innes Seed Compost this was obtained by mixing loam with sand and peat but since good quality loam is not always obtainable other mixtures have been devised which are easier to make up for repeat orders. These soils are used in boxes or shallow pans which have adequate drainage so that they do not become waterlogged. If a seed bed or direct sowing in the final position is being used the soil should be brought as close to the seed compost standard as possible by cultivation and additions. A very small amount of phosphatic fertilizer and lime is usually incorporated but strong fertilizers or rich organic material should not be used in seed beds. If you are buying a seed compost it will be prepared from sterilized ingredients which do not contain any weed seeds and a minimum of micro-organisms which might cause damage to the young seedlings, but for very susceptible seedlings a fungicide suitable for use on seedlings is often watered on soon after germination. Obviously if you are using unsterilized soil you must expect both weeds and more risk of fungal damage. If you sterilize soil yourself follow the instructions carefully. Too much heating can damage the soil structure and even cause inhibition of the growth of some seedlings.

Although exact figures normally have little meaning because of the inter-acting influences of other factors it is still convenient to talk about minimum, optimum and maximum temperatures for seed germination. The ideal aim is to keep somewhere in the optimum region. For many seeds this means about 20°C but for plants of exotic origin like sweet corn and *Cucumis melo* it may be around 30°C. At the other end of the scale are those seeds preferring a lower temperature, like 15°C for annual delphiniums. As a general rule only hardy plants can be sown directly into the soil. Half-hardy annual really need to be started inside or with a little heat, otherwise they will be so late in developing that they will be damaged by frost before they reach their best. One of the advantages of using a seed bed is that the seeds can be germinated before their final site is cleared. Another is that the best spot can be chosen from the point of view of temperature and soil. Most plants can be moved around as young seedlings without too much damage. The exceptions are those plants grown for a single swollen main root where early damage will distort the final growth.

The need for adequate soil aeration has already been discussed in a previous chapter and should be naturally present in a correctly watered seed compost. Many seedlings brought on in boxes or pots are covered with glass to encourage germination. This gives a good humid atmosphere but as soon as the seedlings are up the glass must be raised slightly to promote a little air flow. This will help to deter micro-organisms from spreading through the fragile young plants and also ensure a supply of the oxygen which is needed by the young seedlings and prevent a build-up of excess carbon dioxide although when photosynthesisis starts the amount will obviously be reduced. The slight lowering of humidity around the shoots helps to harden them for their future place in the outside world. The other meaning of 'air' to the gardener is room to grow. Seeds must not be planted too close together or they will become partially etiolated by mutual shading and also compete for soil nutrients. This will produce weak plants.

Once a seed has germinated it needs as much light as possible to promote short sturdy growth. Any paper covering the glass sheet over boxes or pans must be removed as soon as the seedlings appear. Direct sunlight should not be allowed to fall on plants under this type of glass cover or the temperature may well rise to dangerous heights, but when the glass is removed this danger will no longer exist. If it is obviously going to be necessary to thin out or transplant some of the seedlings because the seed was sown too close together, do this as soon as possible. All seeds should be planted thinly—it saves money and time. Small seeds can be mixed with sand or sown from a seed-sower to ensure a sparse sowing. Pelleted seeds are also available which make sowing easier in addition to encouraging the seedling to grow because of the nutrients present in the pellet. The earlier a seedling is planted out in its permanent home the sooner it can become established and really start growing. The later it is transplanted the more roots will have formed and the more damage will result from moving it.

A word about the mechanics of saving one's own seed. Obviously seed from good healthy plants should be selected. Often it may be necessary to control the pollination if you want to be sure of what will come up next year. Let the seed ripen naturally on the plant for as long as possible, enclosing the seed head in a perforated bag if you are afraid of the seeds dropping out. Harvest on a dry day, not when it is damp as this will almost certainly lead to mould growth on the seeds. Let the seeds finally dry off gently in the air before they are put away. Paper envelopes can be used but are best then stored in a polythene bag or closed container to prevent moisture getting at the seeds. Always label seeds with the name and date of collection as soon as possible as memory can be very short.

Chapter Six

Vegetative Propagation

To the gardener vegetative propagation implies perpetuating or increasing the numbers of a plant by any means other than seeds. To the botanist it indicates any form of reproduction in which the sexual processes of fertilization do not occur. These are almost the same thing, but not quite. There are many advantages accruing to the gardener who uses vegetative propagation. Perhaps the most important is that he can be virtually certain that the new plants will be identical to the old ones. Secondly he will have a usable plant in a shorter period of time. Thirdly he will be able to propagate and increase plants which do not readily set seed. Plants that propagate themselves vegetatively also have advantages though these are not quite so obvious. For instance, once the first plant has become established it can spread out from this site to colonize the surrounding area and it will score over the other vegetation because it will be using food and nutrients accumulated by the main plant. Vegetative propagules usually have far more reserves than seeds.

Almost any part of a plant may be used for vegetative propagation but where some specialized organ has been developed for this purpose it is usually given a particular name. Some types of vegetative propagation are a means of enduring adverse conditions, and accordingly possess some form of dormancy mechanism. Others are merely a means of spreading the plant and do not have any special dormancy mechanisms.

The essential requirements for independent life are a root and a shoot system. But since vegetative propagation does not involve an embryo with a preformed root and shoot system roots have to be produced on pieces of shoot, or shoots on pieces of root, or both roots and stems on a piece of leaf, say, as the starting material. Of these various possibilities the formation of new roots on a shoot is by far the commonest even in nature and also the easiest for the gardener. This is probably because shoot buds have very elaborate surface structures and are consequently difficult to produce, while roots can develop inside the tissues where plenty of living cells are available to initiate their formation. New roots which arise from a shoot system and are not derived in any way from the tissues of a plant's original root are known as *adventitious* roots. This does not mean that they must look different or function in a different way from 'true' roots. It merely described the way in which they arise. In the same way buds which arise in positions other than in the axils of leaves are also described as adventitious.

RUNNERS AND RHIZOMES

Strawberry runners are familiar to every gardener. They are lateral stems which grow out from the axils of the leaves on the flowering plant. The first node does not have normal leaves but at the second node a rosette of normal leaves develops. Adventitious roots grow down into the soil and a new runner then starts from the axil of a leaf on this 'plant' and so on.

Some botany books use the term stolon for short runners but others use it more indiscriminately for any more or less horizontally growing stems both above and below ground. The term rhizome is used for the latter type in this book and stolon is not used at all. Rhizomes can be thin and rapidly growing, a feature of great advantage to noxious weeds like field bindweed (*Convolvulus arvensis*). Rather more fleshy ones enable lily-of-the-valley and mint to form dense mats. Taking this trend even further are the thick, stumpy storage organs of Solomon's seal (*Polygonatum multiflorum*) and the flag irises.

There are some garden plants, such as members of the genus *Aster*, in which the rhizomes only last for one season. They are produced in great numbers each year, rotting off in the second year. In this way dense clumps develop. Obviously such clumps should be divided from time to time otherwise the soil becomes completely depleted of nutrients and the stems grow so close that they compete for light and space.

Rhizomes such as those of the flag irises which grow partly exposed should of course be planted at this level and the pieces should be of sufficient size to be able to establish themselves before their reserves are completely used up. This also implies that rhizomes are best moved when full of reserves and not when flowering. Because they grow horizontally it is clearly of no help to a rhizome to plant it pointing vertically upwards. As is usual practice any cuts made to divide up large rhizomes should be clean ones made with a sharp knife because rough bruised areas provide an ideal site for infection by the micro-organisms which abound in the soil.

BULBS AND CORMS

The botanical distinction between bulbs and corms is easily made. Bulbs are made up of enlarged, thickened, bulb-scales borne on a flattened or cone-shaped main stem. Corms do not have fleshy scales but consist of a shortened, swollen stem. The leaf bases of this stem form the fibrous covering of the corm and the fact that they arise at nodes is clear when they are inspected closely and seen to arise on distinct ridges which run around the corm.

In all bulbs and corms the new propagules can only develop fully when supplied with the organic substances formed by the leaves. When the flowering period has just finished the reserves are at their lowest ebb. If the plant is allowed to form seeds the nutrients are diverted away from the young developing bulbs or corms. It is essential therefore both to let the leaves stay on growing for as long as possible and also to remove the flowers as soon as they are dead. In the formation of an onion bulb growth of the roots and stem apical meristem stop and the carbohydrates are passed into the developing bulb scales, which in this particular case are made up of the bases of very young leaves. When an onion sprouts the stored reserves are again mobilized to support the renewed growth of the apex with the consequent shrivelling of the bulb scales.

TUBERS

Tubers can be either swollen stems or swollen roots and are named accordingly. The main difference from the gardening viewpoint is that stem tubers will have many axillary buds on them —the 'eyes' of the potato—while root tubers will normally only have buds on the short piece of stem which is attached to their upper end. As buds are essential for a shoot to develop, the difference in handling required is obvious.

The above different types of swollen storage organs have certain points in common. One, of course, is the fact that when they are increasing in size they draw upon the sugars and other nutrients of the plant very powerfully, draining them out of the other regions and reducing the growth of these other parts of the root and shoot system. Another feature is that the formation of many of them is induced by changes occurring outside the plant. It has been known since the 1920s that the number of hours of daylight which a plant receives can influence the formation of potato tubers. Certain cultivars form tubers much more readily in 'short-days' than during the long days of summer. Such potatoes will therefore produce a crop late in the autumn. Many other aspects of plant growth are also affected by daylength; the effects on flowering are discussed in the next chapter. Daylength of course means the number of hours of light ('day') in the twenty-four hours. It is now known that it is really the length of the night which is important for the plant but because man seems to think naturally about day rather than night the original terms of 'short-day' and 'long-day' are still used. The botanist uses the term *photoperiod* to describe plant responses which are influenced by the length of the day and the night.

When the subject of tuber formation was first studied it was found that potato plants formed tubers more readily in short days but it subsequently became clear that this depends to a large extent on the cultivar. Not only the formation of certain potato tubers but also that of several other storage organs is favoured by short days, among them the above-ground tubers of *Begonia*, the underground stem tubers of Jerusalem artichoke (*Helianthus tuberosus*), and the root tubers of some dahlias. Long days stimulate the formation of some onions, early garlic and various other plants. From experiments it is quite clear that a tuber-forming substance is produced in the plants under the appropriate conditions and that this can move above the plant. In the case of the potato it has been clearly shown that any node, above or below ground, is potentially capable of producing a tuber but normally only a few of the nodes on the underground rhizomes actually swell up.

The daylength is 'sensed' by the leaves and a stimulus passes down the stem, almost certainly in the phloem. If the phloem in a stem is damaged small tubers form just above the damage where tuber-forming substance accumulates. Tuber formation may result from some special combination of auxin and gibberellins which occurs only in a particular daylength or there may be a separate, but as yet unidentified, tuber-forming hormone.

Suppression of vegetative growth can encourage tuber formation and in the potato, pinching back the growing points has this effect. Even on a sprouting potato an incompletely pinched-out eye will produce many small new tubers around it. In some cultivars slow growth caused by low temperatures towards the end of the season can also stimulate tuber formation. Experiments with both natural and synthetic auxins have shown that they can stimulate storage organ formation, but since the storage organs produced by this method are often very small or incompletely developed the use of these chemical does not appear to offer any advantages over letting nature take its usual course. One feature of potato growth which many gardeners may have noticed is that the increase in tuber weight gets more rapid as the tubers grow older. It may take forty days for a potato to reach a weight of two ounces but in the next ten days it could reach half a pound provided, of course, that it was a cultivar which formed large tubers. Those gardeners who insist on digging up their potatoes when the tubers are still young must clearly be prepared to accept a considerable loss in yield. There are of course, certain cultivars which inherently produce a large number of small tubers.

In any plant which forms a lot of separate tubers it is usually found that food materials can be moved from one tuber to another and in this way the number of tubers can even decrease as the plant grows because some shrivel away as the more actively growing ones increase in size. The reserves in tubers can also be mobilized to produce new shoots if an early frost kills off the top growth before the tubers have become fully dormant, which is a good reason for not leaving potatoes in the soil after they are ready for harvest.

Underground storage organs go through a definite process of ripening. 'New' potatoes for example, contain quite a lot of sugar and have thin skins which easily rub off. Old potatoes have thicker, more protective skins and heal more quickly if damaged; the sugar content is much lower and starch forms a great percentage of the reserves. This makes them rather better for commercial frying, quite apart from the economic advantages of waiting till the crop has reached its full weight! Because of the thicker skin old 'ripe' potatoes keep much better than young ones which soon shrivel up. By analogy to seeds it might be thought that the thicker or more impermeable a potato skin is the more dormant the tuber would be. Much scientific study has been devoted to the ways in which tubers can be either kept dormant or encouraged to sprout. Moisture favours sprouting, apparently because it encourages the rapid formation of a corky skin which prevents oxygen reaching the inside of the tuber and so encourages sprouting. As might be expected, storing at a high temperature also encourages growth but it does not follow that the lowest storage temperature is best: in fact, if potatoes are kept close to freezing they start to sweeten, because some of the starchy reserves are changed back into the individual sugar units. For prolonged natural storage the gardener needs that proverbial 'cool dry shed or cellar'. Airtight polythene bags are not suitable for storing potatoes because moisture cannot escape from the bag, and also the oxygen level falls as it is used up by the respiration of the tubers.

Not being content with nature, man has devised various ways of preventing the development of sprouts. It will be remembered that auxins inhibit lateral bud development on shoots and this effect is also shown with potato tubers, which are themselves stems. As little as twenty-five parts of auxin per million of the tuber weight is effective, applied in a mixture with talc. The effect of added auxin is apparently to upset the normal balance between the natural inhibitors and promotors of growth and the auxin is not itself the natural substance which prevents sprouting.

Although it is possible to save one's own 'seed' potatoes, many

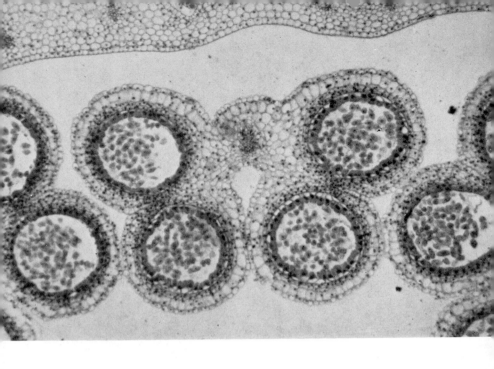

Page 117 (*above*) Section across the developing anthers of *Lilium* showing the four separate pollen sacs in each anther—magnified; (*below*) section of a single pollen sac of *Lilium*: the nucleii and the developing pollen are clearly seen—highly magnified.

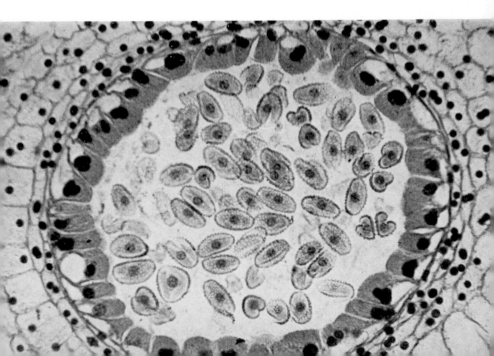

Page 118 (right) A whole carpel from a buttercup flower showing pollen grains on the stigma and the ovule visible as a dark spot inside the ovary—magnified; *(below)* section of a wheat seed to show the embryo with its separate root and shoot already clearly visible—highly magnified.

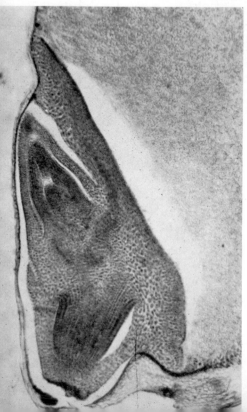

gardeners prefer to buy disease-free, authenticated cultivars each year. If you do save your own there is a danger of diseases being passed on from one year to the next, whereas commercial seed potatoes are grown in areas where there is a minimum chance of them being infected. There is no advantage in selecting the biggest potatoes in the hope of getting a crop of large tubers because the average size of tubers is determined by the cultivar and any extra-large ones only reflect some chance environmental conditions in that particular plant which will not be inherited. It is possible to increase the number of plants which can be obtained from one tuber by cutting it into many pieces, each containing at least one eye. Naturally the cut surfaces should be allowed to dry off before placing the pieces in the soil.

Hundreds of chemicals have been tested for their abilites to break bud dormancy. Ethylene chlorhydrin and various thiocyanate salts can be used in solution at a strength of less than one per cent to 'force' tubers. Using them it is possible to grow a new crop of tubers before the untreated, dormant, ones have even started to sprout—provided you have the right growing conditions for them once the sprouts have started. Because of the high cost this technique is mostly used by botanists to check on the quality of 'seed' tubers or for breeding new cultivars, rather than by the agriculturalist for food production. With these chemicals the natural correlation of the eyes is maintained and only one bud normally grows out from an eye. With other chemicals, such as thiourea, several buds may grow out from a single eye and this may be an advantage for producing seed potatoes if they would otherwise grow too large under natural conditions. Because many stems grow out from the treated tubers a greater number of smaller seed potatoes will be formed, which purchasers prefer to large ones which have to be cut before planting.

ROOT AND STEM CUTTINGS

Now we must turn our attention towards the more artificial ways

H

of propagating plants. Perhaps the easiest, and the one which most gardeners have tried at one time or another, is propagation by cuttings.

For almost every stem cutting a piece with short internodes is best because it will contain a maximum number of vegetative buds above ground and a larger number of nodes below ground, which may encourage rooting. Normally the apical bud is left on the cutting unless a very bushy plant is required because it will exert its 'apical dominance' to encourage the development of a good strong main stem. The portion to be buried in the soil is stripped of leaves which would rot away and encourage micro-organisms to attack the cutting. The above-ground leaves may need to be reduced somewhat in number and certainly any which will actually touch the soil must be removed before they rot.

In some plants it is possible to take root cuttings. Plenty of garden flowers such as *Gaillardia, Phlox* and oriental poppy can be propagated in this way. These cuttings need to be a reasonable size, perhaps a couple of inches long and buried not more than about an inch below the soil. When taking root cuttings experienced gardeners usually cut the top end straight across and the lower end at an angle so that they can be sure of planting them the right way up. Larger thin pieces may be planted horizontally rather than vertically. Suckers are a form of natural root 'cutting' and can be reared to full size if this is desirable. Layering is an extension of the stem cutting method of propagation and is of particular value with plants that are difficult to 'strike' by other means. A slit is made in the underside of a stem, wedged slightly open if there is a tendency for it to spring back together tightly, and the cut portion is then buried under an inch or two of soil. Pegs or stones may be used to hold it down. What is known as a serpentine layer has two or three buried regions along its length. In the aerial layer, a method used on shoots which cannot be brought down to the ground, a false 'soil' of rooting medium is made up into a ball around the cut region. The advantage of

120

layering is that the 'cutting' can draw upon the parent plant for water and food to tide it over the critical days before it becomes large enough to be self supporting and is severed from the parent. This is especially useful if you cannot provide ideal rooting conditions for cuttings.

What of the botany behind the taking of cuttings? Firstly all parts of a plant show a phenomenon known as *polarity*. This means that even if a stem or root is cut out without a growing apex on it the two ends of it are different. Roots will tend to form on the end furthest from the stem apices or nearer to the root apices. If you suspend a cutting upside down in suitable conditions roots will still come from the 'root' end and the shoots from the other one but this is only of value as an experimental demonstration because under gardening conditions such plants would quickly die.

The polarity is apparently caused by a gradient of hormones in the cutting. Auxin appears to be concentrated at the original lower end, probably because it normally only moves away from the apex it will tend to build up at a cut surface. At the upper end cytokinins are present and the development of vegetative buds is favoured by these hormones.

With the development of synthetic auxins the gardener now has a powerful ally in his attempts to propagate. The natural auxin, IAA, is rarely used because it is not very stable in the soil and soon breaks down into substances which have no root-promoting properties. Literally hundreds of chemicals have been tested, in all sorts of combinations and with various methods of application. The number which are actually used is much more limited, by effectiveness, cost and for various other reasons. Commonest are NAA (naphthalene acetic acid) and IBA (indole butyric acid) and combinations of the two. The auxins can be mixed with an inert substance like talc, or dissolved. In either case the end to be rooted is dipped into the auxin preparation in the prescribed manner and then planted in the normal way.

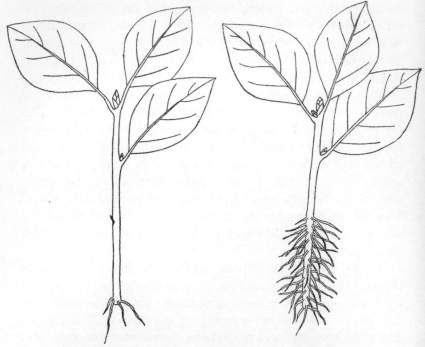

*Auxins can have a dramatic effect in stimulating the
rooting of cuttings*

Actually auxim stimulates the production of new roots (see
above) but it inhibits their subsequent growth. It is therefore
of no advantage to water the ground with auxin preparation or
fill the hole with it before you insert the cutting. This will do
more harm than good. The growing roots draw upon the reserves
in the cutting and it has been found that in a few plants there is a
shortage of vitamins, or at least better rooting is obtained if
vitamin B_1 is added. The presence of food reserves may explain
why some cuttings are best taken later in the year when they
have built up a stock of the products of photosynthesis. Flowering shoots are rarely satisfactory for cuttings, probably because

122

of the competition of the flowers for foods and if you must use them remove the flowers first.

The most serious danger with cuttings is that they will dry out and die. A root system takes some time to develop and it is often months before it is big enough to supply the cutting with adequate water. Even when a cutting is sprouting, it may still have a very small root system so that young cuttings should not be moved until rooting is well advanced. This is particularly so with woody cuttings taken in the autumn which may not be fully rooted until well into the spring. During the time when the root system is not fully developed steps must be taken to cut down water loss from the shoot. Cuttings stuck into bare ground in the full sun are doomed. If they must be planted directly in the soil choose a cool shady spot—it doesn't matter if they cannot be left there once they are rooted. The soil needs to be kept moist but not waterlogged—the same conditions as are required for the early growth of seedling roots which means a well drained but water-holding soil. Actually the 'soil' part is not too important at first and cuttings can be rooted in sphagnum moss or other damp fibrous material. They will, however, need transferring when they have developed some roots so that it may be better to obtain the correct soil mixture rather than risk damage during trans-planting.

LEAF CUTTINGS

With some plants there is not enough stem to take cuttings from. In others experience has shown that some other part can be used for propagation. Commercially there may not be enough shoots to give the desired number of plants, particularly if the plant is a new cultivar. One or other of these reasons leads to the use of other types of cutting but as the pieces used get smaller and more delicate so the techniques needed become more complex. Most gardeners have heard of leaf cuttings, which can be used for many species of plant but chiefly those which have rather fleshy leaves or are genuine succulents. The petiole (leaf stalk) end can be inserted into the rooting medium rather like a stem. With plants like *Begonia rex* cuts can be made in the lower side of the main veins and the leaf pegged down. In this way several plants can be formed from one leaf. It is interesting that leaves used as cuttings often live much longer and even grow larger than they do if left on the parent plant. Presumably their growth is inhibited on the whole plant by the action of the other leaves and stems.

The rooting of pieces of leaf is apparently dependent on an interplay of auxins and cytokinins just as in stem cuttings. Leaves are sites of production of auxins and experiments with thousands of different species have shown that most of them will naturally form roots if properly looked after. Many do not naturally form buds however, presumably because of a lack of cytokinins and this fact led naturally to experiments with added cytokinins. Despite the high cost of these substances there is some advantage in using them on expensive plants where stem cuttings are not satisfactory or do not give a sufficiently rapid rate of increase; they have for example increased bud formation on leaf cuttings of a number of greenhouse plants such as *Saintpaulia ionantha* and various *Begonia* species. The use of cytokinins is not quite as easy as this might suggest because root formation tends to be inhibited and since both roots and buds are needed for a complete plantlet skill may be needed in the application of cytokinins. In *Peperomia sandersii*, for instance, bud formation seems to be entirely dependent on prior formation of roots and cytokinins inhibit rather than promote plantlet development.

TISSUE CULTURE

Although not really of any practical significance, the rooting of all manner of bits and pieces taken from plants has been reported by botanists from time to time. Some fifty or sixty examples are available of different inflorescences which have been treated as cuttings and grown roots. At the same time there is a conversion of at least some of the inflorescence 'back' (if that is the correct term) to vegetative growth. Cactus flowers, bean pods, even isolated petals have all been induced to form roots when cultured under the right conditions. The smaller the pieces the more they have to be cosseted and supplied with various sorts of foodstuff. This leads us naturally on to a consideration of the use of *tissue culture*. In this technique small pieces of tissue are taken from almost any part of a plant, freed from any micro-organisms which would damage them, and transferred to sterile growing media in sterile containers. Since the small pieces of tissue are not self-supporting—indeed they may not be green or have any particular structure at all—they must be supplied with sugars, mineral elements and the correct concentrations of various plant hormones.

At first tissue culture was only of academic interest because the tiny pieces just grew into irregular lumps of callus tissue. When the various recently discovered plant hormones were added, at the correct concentrations, the development of roots, buds and finally complete plants could be obtained. If these were carefully handled they could eventually be planted out into soil and grown on into mature plants. Now the scientific gardeners really had something to play with. Pieces of pith, leaf, root, or anything else with living cells, can all be cultured. The easiest pieces, however, are from those parts which already have cells which are dividing and growing. The apices of stems have been used extensively, giving rise to the technique of *meristem culture*. A tiny piece from the tip of any growing branch is all that is needed. One particular advantage which has already been exploited is that the young apices are frequently free from virus, even if the rest of the plant is infected. Virus-free stock can be built up in this way. Large numbers of particularly valuable plants can also be raised by encouraging the parent to become

125

bushy, perhaps by pinching out, and then taking the apices for meristem culture (see diagram below).

Even meristem culture is not the ultimate in vegetative propagation. The next question is what is the smallest amount of tissue which needs to be taken to give a complete plant? A plant is made up of hundreds of thousands of cells. Can a single cell give rise to a complete plant? Certainly a single fertilized egg-cell can, but what about a cell taken from a leaf or piece of stem? It has now been definitely proved that a single cell contains all the information necessary to grow into a complete plant. A piece of plant tissue, say a developing embryo or piece of petiole, is broken up into its individual cells. These are then spread out, isolated from one another, on a suitable growth medium and each develops into a complete plantlet. These are known as *vegetative*

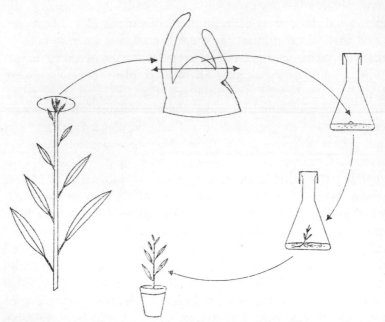

A diagram to illustrate the technique of apical meristem culture

embryoids to distinguish them from the true embryos found in seeds. The embryoids all resemble the parent plant exactly because they do not result from the fertilization of an egg cell.

BUDDING AND GRAFTING

These constitute a distinct type of vegetative propagation. Although there are plenty of examples of natural grafting of tree roots to one another and also of stems in very close contact fusing together, the joining of part of one plant to another plant is mostly the result of the gardener's art. The principle of all the numerous techniques is the same. A stock is taken which has an active root system and a certain amount of stem and onto this is attached a piece of some other plant, known as the scion if it has a stem, or a 'bud' if that is all it is. The two are held together in intimate contact by supporting straps until the tissues fuse to give a strong joint, when the straps can then be removed. Excess stock is trimmed away at an early stage to encourage the graft union to form and the new shoot to grow strongly. In a way the technique is rather like tissue culture except that we use a living plant to supply the needs of the desired plant instead of a complex synthetic medium. The stock remains as part of the new combined plant and the desired cultivar does not have to produce its own roots.

All manner of advantages come from the use of budding and grafting. It is a very rapid way of obtaining large numbers of a selected cultivar. Even two or three buds of the desired plant can be turned into a respectable shrub in a couple of years: rose growing provides the classical example of this. Another advantage is that poorly rooting cultivars can be grafted onto the strongly growing roots of another less desirable cultivar. In many cases it is even possible to graft onto a completely different species. This again is common practice in rose growing. Although the shoots produced by budding and grafting are genetically identical to the donor cultivar it has been found that the type of root stock can influence the development of the scion in some cases. Various types of apple rootstock have been developed, many at the East Malling Research Station, which can affect the size of the final tree. The fruit produced is still, however, usually true to the donor variety.

There is no theoretical limit to the number of different cultivars which can be grafted onto one stock. Practical, aesthetic and economic factors control the number. Where more than one cultivar is grafted due consideration must be given to the growth rates of the various donors otherwise a lop-sided and difficult-to-manage plant will result. One technique of multiple grafting, with definite advantages, which does not appear to have reached many gardeners is the placing of cross-pollination cultivars onto one plant. In this way even the small garden can have success with only a single specimen of a plant which requires cross pollination before it will set fruit; cherries, for example, have very stringent requirements for cross-pollination.

The choice of stock and scion is firstly determined by experience. Those combinations which give a good union may subsequently be developed commercially where the availability of stock and cultural factors usually limit the final choice. Fortunately for the gardener plants do not have the same immunity problems which arise when animal tissues are grafted together. With animals only pieces of a single individual can usually be grafted together successfully and pieces from other individuals, even if the same species, do not combine well. This is because the animal reacts to the piece of tissue being grafted and 'rejects' it. With plant grafting, tissues of the same cultivar are normally fully compatible—indeed they should be as they are as similar as true identical twins. Different cultivars of the same species can almost all be grafted together. Unions between different species are less common, between different genera rare and between different families very rare.

To obtain a good graft union the cells of stock and scion have to fuse together. The question of polarity of cuttings has already been discussed (page 121) and the same principle applies to grafting. If a piece of stem or a bud is placed upside down onto the stock there is considerable difficulty in obtaining a union. In the correct alignment the movement of substances in the translocation (conducting) system of the stock and scion would be in the same direction. When the scion is inverted its vascular system, as far as can be ascertained, continues functioning in its original direction and thereby upsets the flow of foods across the graft. Detailed studies of the vascular supply to upside-down buds indicate that considerable twisting of the vascular systems may occur so that the ends can join up with the normal polarity.

In woody plants the fusion of stock and scion is very much affected by the cambial layer in the stem. The vascular cambium is a layer of cells which are meristematic (normally show division) and which produce new xylem on the inside and phloem to the outside. This is the way in which woody plants increase in diameter, adding to the xylem ('wood') each year. The phloem is structurally a much weaker tissue and gets crushed against the restricting cork around the outside of the stem so that it does not increase much in thickness from year to year. The cells of the cambium do not appear to have any polarity and fuse easily to other cambial cells, presumably because the cell walls are not fully formed and are therefore still capable of growing firmly together. The art of grafting in woody plants is to ensure a maximum of contact between the cambial regions of the stock and scion or bud. If the gardener had to perform a delicate operation to dissect out the cambium the whole practice of grafting would probably never have arisen. Fortunately the cambium provides a line of weakness between the wood and the other layers on the outside of the stem. When we peel off the bark of a stem the moist, glistening cambial region is revealed, partly attached to the bark and partly to the xylem, and by bringing the cambium of the stock close to that of the scion or bud we have the makings of a good union. After the parts have been bound in place it should only be a matter of time before the cells fuse together firmly.

Despite the great importance of the cambium in woody plants it is still possible to obtain grafts between plants which do not have a cambium. The problem with these is providing support and preventing the wilting of the scion. It is even possible to graft woody plants onto herbaceous ones, and vice versa. One often-quoted example is the linking of *Sophora japonica* to the ordinary bean (*Phaseolus vulgaris*).

All manner of substances and influences have been shown to pass across a graft union. It is only to be expected that inorganic and organic nutrients and hormones will move in the xylem and phloem systems once they are joined together, and in this way the interaction can bring about changes in overall plant size and habit and in the size of leaves and fruits. The flowering and tuber-forming responses can migrate through a graft, and this is usually taken as evidence of their hormonal nature. Less tangible characters such as life span and resistance to disease can also be affected by grafting, but there appears to be very little evidence of genuine changes in the genetical nature of the stock or scion. The exceptional effects found in chimeras are discussed in the final chapter of this book.

Chapter Seven

Flowering and Fruiting

In previous chapters much attention has been paid to the apical meristems of the shoot, which in the vegetative condition produce the leaf primordia and the stems. Flowers are normally considered by botanists to be greatly modified stems. During the conversion of a vegetative apex into a flowering apex, drastic reorganization occurs. Instead of leaves the apex produces the various parts of the flower (see diagram over). It may even split up into a number of separate apices, each giving one flower of an inflorescence.

What causes flowering to start? Once a vegetative apex has been converted into a reproductive (flowering) apex the actual development of the fully opened flowers is merely a matter of time. Only in a few rare cases, such as the cuttings made from inflorescences mentioned in the last chapter, does a reproductive apex change back into a vegetative one. It is as if a whole new sequence of development has been switched on. The reproductive apex need not develop immediately into a flower but can remain 'dormant' until some other influence causes it to actually develop into a flower. (See photographs on page 84.)

Quite apart from differences between plants in the actual

131

structure of the flowers, there are also great differences in the conditions which are necessary to promote flowering. Some plants flower in every month of the year while others do so only in a very restricted season. Botanists have concentrated their attention on this later group of plants because they are the ones which are easiest to study. They are also the ones which give most difficulty to the gardener. If a flower is seasonal we can examine the climatic conditions prior to and during that season to obtain information about the factors that induce flowering. Even in those few plants where flowering can be linked to some definite environmental stimulus there are plenty of complications. Different cultivars of the same plant may respond in totally different ways. To the gardener this can be an advantage, giving a spread

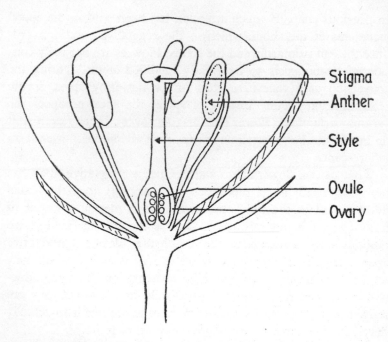

— Stigma
— Anther
— Style
— Ovule
— Ovary

Parts of the flower involved in seed production

132

of flowers throughout the year, but to the botanist trying to study flowering it has meant that his work is usually limited to one particular cultivar of one particular species. When a gardener looks at botany books dealing with flowering he will keep meeting the same plants, simply because these are the ones which have been studied. He may never come across the actual plant he is interested in because it has not been examined by botanists. Of course the gardening books tell us when plants will flower under natural conditions. The whole point of looking in other books is to try and find out how to make them flower in unnatural conditions or exactly when we want them to. But at least the gardener delving into botany will gain an insight into the factors which control flowering and then, carefully observing his own plants, he can try to manipulate conditions to promote flowering. One major problem for the gardener is that his control of the environment is usually much less accurate than that of the botanist who studies flowering scientifically and has very accurately controlled growth chambers available for his plants. All this is really a warning against expecting too much, but with intelligent application the average gardener can expect to gain greater control over his plants and even get some to flower at unusual times of the year.

THE CONTROL OF FLOWERING

One of the most important environmental factors controlling production of flowers is the length of the day. This has already been mentioned in connection with the formation of storage organs (page 113), but the subject has been even more extensively studied with respect to flowering. Those species which are strictly controlled by daylength are able to synchronize the flowering of individual plants exactly and this provides a maximum likelihood of cross-pollination. Plants can be separated into three simple groups with respect to flowering (see diagram over).

Short-day (SD) plants flower in short days, *long-day* (LD) plants in long days, and *day-neutral* plants flower in any daylength. Within each group the complexities become greater the closer one looks. Within the two groups which are affected by daylength there are some plants which will only flower if given more or less (as the case may be) than a certain number of hours of light. These are called *qualitative* (absolute) LD and SD plants respectively. Those plants which are merely encouraged (promoted) in their flowering by particular daylengths are described as *quantitative* LD or SD plants. In this group flowering can occur in any daylength but will be better if the plant is grown under the appropriate lighting conditions.

The other major environmental factor which interacts with a daylength response is temperature. Plants from temperate

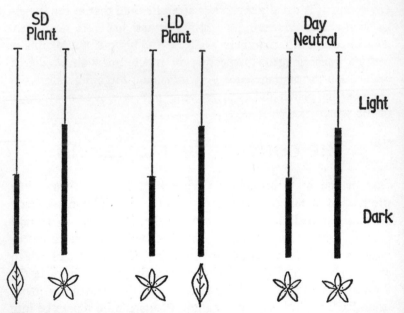

The three types of flowering response with respect to daylength

Page 135 (above) Fallen apples covered with the spores of the brown rot fungus—these can act as a source of infection for next season and must be cleaned up; (*below*) a tiny part of a saprophytic fungus colony showing a few of the millions of spores which can be produced—highly magnified.

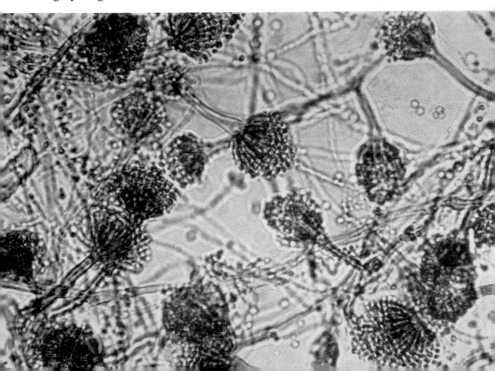

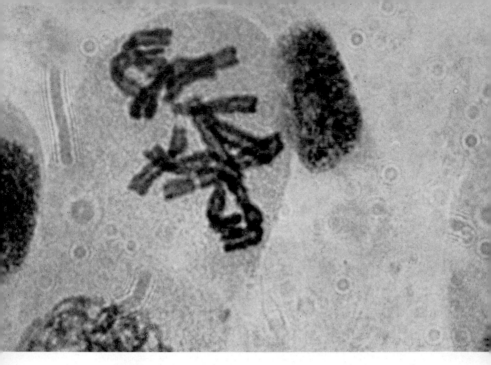

Page 136 (*above*) The chromosomes of the broad bean just prior to separation during cell division—highly magnified; (*below*) the chromosomes of the broad bean separating during cell division: a new cell wall will form between the two sets as they return to the 'resting nucleus' state—highly magnified.

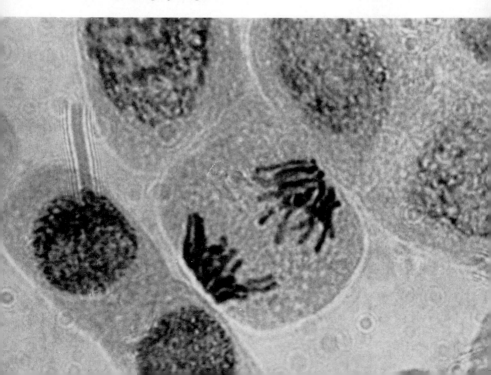

climates may require a fairly lengthy period of low temperature, simulating a normal winter, before flowering, whereas plants brought from warmer climates may require a particularly high temperature. Others again may prefer an alternation of two different temperatures. The complexities of flowering arise from the interactions between these different requirements. There are literally hundreds of different combinations. This is one reason why so many gardening books deal specifically with the culture of a particular type of plant. These are based on actual observations of the plants grown under various conditions. If we can follow their instructions we shall succeed in growing the plant satisfactorily but if we cannot do so or want to produce flowers at unusual times other information is needed.

EFFECTS OF DAYLENGTH

As we mentioned previously plants can sense the length of the day and they do this by means of a pigment called phytochrome. The interesting thing about phytochrome is that it can exist in two different forms. One form known as P_r absorbs red light and in so doing becomes converted into another form (P_{fr}) which absorbs far-red light. Far-red light is just beyond the range of the human eye's ability to see. Sunlight contains both red and far-red light. In the dark the P_{fr} is converted back to P_r or changed into a completely different substance. This provides a simple way in which the plant can sense whether or not it is in the light: if some of the phytochrome is in the form P_{fr} it must be in the light. It is the P_{fr} which actually does things in the plant and it is able, in some way not yet clearly understood, to switch on the synthesis of new substances such as red pigments, to cause new structures (hairs for example) to grow, and also to switch off other processes such as stem elongation. The gardener with a packet of mustard seed can easily demonstrate these effects for himself if he plants some in normal day and night conditions

137

J

and some in continuous darkness. Botanists are particularly interested in phytochrome because it appears to be the trigger which sets off many of the growth effects caused by light. Instead of using darkness, which takes several hours to reconvert P_{fr} back to P_r botanists often use far-red light which brings about the change very rapidly because it is absorbed by the P_{fr}.

Although it is the apical meristem which actually changes during the induction of flowering the *photoperiods* (daylengths) are sensed by phytochrome in the leaves. In some plants it is only necessary to subject a single leaf to the correct conditions in order to obtain flowering. Obviously some stimulus moves from the leaves to the apices but since plants do not possess a nervous system of the type found in animals stimuli appear to move around plants mostly as chemical substances by way of the vascular system, particularly the phloem. Phytochrome itself does not appear to be the chemical messenger (hormone) responsible for flowering but an unidentified substance known as floringen is produced in the leaves after they have received the correct light regime. This hormone then moves out of the leaf, accompanying the organic substances produced by photosynthesis, by way of the phloem. It may take a couple of days for enough of it to move out of the leaves to induce the vegetative apices to change into reproductive ones.

In some plants (the strawberry is a good example) there are also inhibitors of flowering which are produced during the photoperiods that do not promote flowering. Recent research indicates that such inhibitors may be more common than was previously thought to be the case. So for flowering there can be separate stop and start signals and it may be the relative strength of these signals which decides whether or not a plant will actually flower.

When certain plants are placed in conditions known to promote flowering they still may not flower and there appears to be a rather ill-defined state called 'ripeness to flower' which a plant must reach before it can respond. This may be due to a requirement for some other treatment such as a period of low temperature, prior to the correct photoperiods. In other plants it is more a question of age. The cotyledons of the SD plant morning glory will respond to short days, but it is much more typical of other plants to find that several foliage leaves must be produced before a response is obtained. Going to the full extreme some trees and other perennials may not flower until they are many years old. The juvenile climbing form of ivy, with its five-lobed leaves, is completely different from the flowering form with entire leaves. There is some evidence that the food reserves built up by the plant must reach a certain size before flowering occurs. Some interesting experiments were carried out on birch trees, in which it was found they could be made to flower when only a year or two old if kept growing continuously in a warm greenhouse. The alternative of growing them in artificially produced, slightly shorter than normal, 'years' with an alternation of summer and winter conditions did not speed up flowering.

EFFECTS OF TEMPERATURE

Something more must now be said about the low temperature promotion of flowering or vernalization. Many plants can be made to flower earlier by subjecting the young plant to a period of chilling before the photoperiods which promote flowering. A great deal of the early work on this aspect of flowering was carried out on cereals, particularly wheat. It was found that winter wheat, which normally has to be planted in the autumn to flower the next summer could be made to flower in a much shorter period of time provided the young seedling, or even the damp seed, was subjected to a week or so of low temperature. Some botanists use the term vernalization in a wider sense to cover any form of growth promotion caused by low temperatures, such as the stimulation of germination which has already been mentioned (see page 102).

The temperatures which are effective in vernalization vary in different species but are usually somewhere between 0°C and 10°C. Commonly 5°-6°C is used. The period of exposure to chilling also varies according to the species. Some may only need a week or two but others may benefit from two or three months. It is found that if the low temperature is followed immediately by rather high temperatures the plants become 'devernalized' and will not flower unless chilled again. Although cereals can be vernalized as damp seed or very young seedlings, most plants must reach a certain size before they will respond. Many biennial plants have a chilling requirement for induction of flowering, which obviously simulates the conditions met in winter. Often they grow as rosettes in the first year and can be vernalized in this state.

Many but by no means all plants with storage organs such as bulbs and rhizomes require a chilling period before they will flower. Some initiate flowers before the 'rest' period and do not need chilling, and others form them after being planted out in the spring. Only those which lay down their flowers during the winter (chilling) period are promoted by low temperatures. Although the flowers may be induced by chilling it often needs a quite complicated sequence of different temperatures to obtain flowering stalks of the correct length and the proper opening of the flowers. Commercial firms who specialize in obtaining flowers at the 'wrong' season have to watch these points very carefully.

The growth of many perennial plants, especially trees and shrubs is influenced greatly by the daylength and the low temperatures of winter. In many the short days of autumn bring on the dormant state, accompanied in deciduous plants by the abscission of the leaves (leaf-fall—see plate on page 83). A period of low temperature is often needed before the dormancy can be broken by exposure to favourable conditions. Most flower arrangers know that it is no use bringing in certain shoots before

Christmas because they will not open their buds, but that after a reasonable exposure to the low winter temperatures many can be made to open just by bringing them into the warm. The dormancy in other plants can be broken by exposure to long days and occasionally by really high temperatures (30°-40°C) for some time.

Although he can buy his flowers out of season from the florist or use pre-treated bulbs for early flowering it is quite possible for the gardener who has a little refrigerator space available to manipulate the flowering of his own plants. If they are wrapped in a polythene bag or in a little soil and put for a few weeks in the outer compartment, this can work wonders. The main risk is that he will become too enthusiastic and push all manner of plants into the refrigerator without first checking that they will benefit from the treatment! Daylength can be extended by artificial lighting and this may be quite cheap because it does not have to be particularly bright. To be effective it must however be regularly applied so that a time switch is almost essential unless your memory is infallible. A slightly cheaper alternative is not to extend the day but instead to use a 'light-break' in the middle of the night. By far the simplest method of adjusting daylength is to use light-tight covers to reduce the length of the light period. This technique is used commercially to obtain early flowering of short-day plants.

CULTURE AND CHEMICALS

In many plants there is an inverse relationship between vegetative growth and flowering. At one time it was even thought that flowering was entirely due to nutritional factors: plentiful supplies of nutrients, such as nitrogen, causing vegetative growth and more adverse conditions favouring flowering. This simple theory is no longer accepted as such but it is nevertheless possible to induce flowering by methods which upset normal growth.

Ringing the bark of branches and stems, tying branches into a downward pointing direction, pruning or pinching out terminal buds, have all been used by gardeners to encourage flowering in certain plants. A period of drought can sometimes promote flowering. Although in many plants flowering is associated with a slowing or stoppage of vegetative growth this is not necessarily so in every plant and even different cultivars can respond in different ways. Late cultivars of tomato, for example, can be made to flower earlier by removal of young, expanding leaves but this treatment has no effect on early cultivars. Of course, these drastic treatments do not always result in goood yields of fruit because an adequate nutrition is needed for full fruit development,

In certain plants it has proved possible to induce flowers by the use of chemicals and hormones but it is not thought that any of those which are currently in use are the natural flowering hormone. It is well known that gibberellins can cause the bolting (flowering) of some plants and that where a requirement for a low temperature treatment exists gibberellins can be used in place of the chilling. In plants without a chilling requirement these same hormones can sometimes substitute for a long-day requirement but never, apparently, for a short-day requirement. Gibberellins promote stem growth and the plants most affected by them are those with very short stems which grow as rosettes (see opposite). In some cases the stem can be made to elongate but it does not bear flowers. For this and other reasons gibberellins are not thought to be identical with florigens. Auxins can promote flowering in a few plants; when they are used on pineapple and litchi vegetative growth is inhibited and flowering promoted. Botanists have prepared extracts from plants in flower which can promote flowering when applied to other plants, even of different species. The active ingredient has not yet been fully identified but a great deal of experimental work is being carried out at the time of writing this book. Often this type of work starts with a whole field of flowering plants and ends up with a few drops of chemical in a tube! Incidentally the flowering response can be passed by graft from one plant to another and this has occasionally been used to encourage flowering in cultivars or species which are very intractable to other manipulations. The results of grating experiments suggest that the same flowering substance is produced by SD plants and LD plants, so there is still hope of finding a simple substance which will promote flowering in a whole range of different plants.

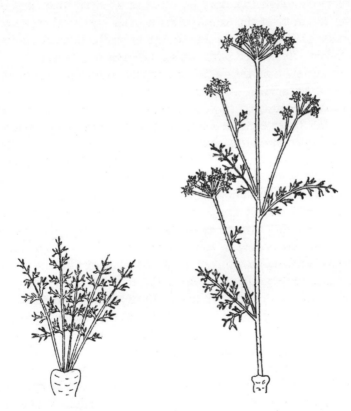

The effect of adding gibberellins to a vegetative rosette

143

POLLINATION

In the world of nature flowers are produced as an aid to reproduction by seeds. An almost indescribable variety of sizes, shapes and colours exist but certain basic types of flower can be distinguished and these are used by botanists when classifying and identifying plants. Even the amateur gardener can soon learn to recognize most members of the common families. Many flowers are *hermaphrodite* (bisexual) and have both ovaries and stamens although this does not necessarily mean that they both reach maturity at the same time. In other species there may be separate male and female flowers, either on the same or on different plants. Associated with the sexual parts of a flower are other structures such as the petals and sepals, although both of these may not be present or distinguishable. It is generally accepted that insect-pollinated flowers existed before the wind-pollinated and other types evolved (excluding conifers and their relatives). Many flowers are attractive to insects by virtue of their colours, shapes and smells. Wind-pollinated flowers tend to produce masses of pollen, to have large receptive stigmas for it to blow onto and a great reduction in the 'attractive' parts of the flower.

Once initiation of the flowers has occurred there seems little doubt that the various plant hormones, but particularly the auxins, are concerned in subsequent flower development. As the flower opens, its growth rate slows down markedly and so does its auxin content. In most cases the stimulus of pollination is needed before growth is resumed. In normal circumstances pollen does not germinate until it falls on a receptive stigma but botanists have been able to grow and study many types in laboratory culture. One very interesting finding is that boron stimulates both pollen germination and its subsequent growth. Calcium also stimulates pollen tube growth and in some flowers it appears to be a gradually increasing concentration of calcium which leads the pollen tube down the style and into the ovary. In other plants

144

this growth is apparently influenced by other 'attractive' substances. In some plants there are also substances present which inhibit the growth of certain pollens. Often pollen will only germinate on the stigma of the same species of plant. Even with a single species the pollen from different individuals may grow at different rates. Since fertilization does not occur until the pollen tube has grown into the ovary, contacted the ovules and grown close to the egg cell the different rates of growth of pollens can control which type will fertilize the egg cell. In many species the pollen from other individuals or cultivars germinates more quickly or grows faster than that from the anthers of the same flower. This is part of the problem of incompatibility which occurs in some plants, such as the cherry where only certain kinds of pollen will produce fruit. There are structural features in many flowers which can also favour cross-pollination between flowers. The genetical aspects of incompatibility are complex and often two or more 'types' of a plant exist. If cross-pollination does not take place some plants have mechanisms by which the pollen from the flower's own anthers is deposited on the stigmas. Often this involves the growth of some part of the flower so that the stigmas and anthers are brought closer together. One feature of pollination which is of interest to the gardener is that large masses of pollen on a stigma cause a mutual stimulation of growth with consequent beneficial results in seed and fruit production. If natural pollinators are not plentiful the gardener can often help by transferring pollen himself to supplement the few grains which may already be on the stigmas. To complicate the problem even more the stigmas of some species are receptive to pollen for only a few hours rather than for several days.

Successful pollination is usually followed by fruit set and the abscission of the petals and in some cases, the stamens as well. The mere act of pollination or the germination of the pollen may be sufficient stimulus to start fruit growth even before fertilization has occurred. When a large amount of pollen is applied to a

145

stigma the subsequent fruit growth is often greater even if the pollen used is from a different species. When the gardener is transferring pollen which he only has a little of, some advantage can be gained by mixing it with a 'carrier' pollen to ensure adequate fruit set. This other pollen has to be from a totally different species of plant so that it is unable to cause fertilization or the carefully arranged cross will be invalidated.

FRUIT SET

Pollen contains auxins and botanists soon found that fruit development could be obtained in some plants without the need for pollination. Any fruit which develops without fertilization is called *parthenocarpic* and it will not contain any seeds. There are some cultivars of edible fruit which naturally produce seedless fruits and these are normally found to contain a higher concentration of natural auxins than seeded cultivars of the same species. This natural parthenocarpy occurs particularly in species which have large numbers of ovules per potential fruit such as bananas. pineapples, tomatoes, melons and figs, although in some cases there may be a mixture of seeded and seedless fruits. Seedless fruits can arise not only from a failure of fertilization but also from an abortion of embryos after fertilization.

Although applied synthetic auxins are successful in obtaining parthenocarpic fruit set in peppers, tomatoes, eggplants, holly, figs and many members of the cucumber family, plus a few more, most fruits cannot be set with auxins. The range of plants, responding to a 'chemical' set can be increased by using gibberellins which are effective in plants such as apple, pear, cherry, apricot, peach, grape and others. The cytokinins do not appear to be useful for fruit set at the present, but they are almost certainly involved in natural fruit development.

Natural fruit set is limited by many different factors. Even when many flowers are produced pollination may fail due to lack of compatible pollen. Climatic variations of temperature, light and rain can also influence pollination adversely. Rain sometimes appears to cause damage by swelling the pollen up until it bursts, though the liquid on stigmas usually contains a large amount of dissolved sugars which prevent this bursting. Competition for nutrients and correlation effects (see page 81) also influence the number of fruit set. When some fruits are already developing they normally tend to inhibit the production of further flowers and fruits. Hence the advantages of the common gardening practice of continually picking off dead flowers or harvesting pods and fruits as soon as they are ready.

FRUIT GROWTH

After fertilization the embryo and associated tissues and structures of the future seed start to develop; these have already been discussed in Chapter 5. Most of our knowledge of the development of structures outside the seed comes from studies on the commercially important fleshy fruits. In some fruits such as the apple the rate of growth of the flesh gradually increases to a peak and then slows down again. In others such as the late cherries and most stone fruits there are two periods of rapid growth separated by a period of slow growth. Although it is known that auxins are involved in fruit growth there are a lot of other substances, many incompletely studied, which also affect the rate of growth, and parthenocarpic fruits set by hormones often lack the characteristic shape of natural fruits containing seeds. Pears produced by artificial auxin may be oval or cylindrical rather than 'pear-shaped'. Where some of the seeds in a fruit fail to develop this often results in a lopsided fruit because developing seeds produce many growth 'factors'. A simple example occurs with strawberry seeds which produce auxins and if some of the seeds are removed at an early stage of growth that part of the fruit does not naturally swell up but can be made to do so by adding auxin.

In fruits like the apple, for any one cultivar there is a relationship between the size of natural fruits and the number of seeds in the fruit; more seeds are present in the larger fruits. The rate of growth of seeds and fruits are not, however, necessarily the same. In beans the pods grow mainly before the seeds, while in fruits such as the cherry which show two rapid periods of growth the slow period in the middle is the time when the seeds are developing most rapidly. Interestingly the Early Richmond cherry has embryos which abort before this period of seed growth starts and in consequence the fruit goes on growing without slowing down and so becomes mature at an earlier time.

Many fruits show a variation in the amount of auxins and other growth promoting and inhibiting substances as they develop. This is sometimes associated with periods when many of the fruits fall off the trees; in apples 'early drop', which occurs just after the ovary starts to swell, and 'June drop', which occurs later during the period of rapid embryo growth, can be prevented by spraying with synthetic auxins. Commercial growers sometimes use auxin to even out the crop variations from year to year. There is, however, in many plants, a relationship between the number of fruits per plant and the size of the fruits, presumably due to competition for nutrients, although this competition may well be controlled by hormones produced by the developing fruits.

The swelling of fruits is sometimes due to new cells being formed and sometimes due to enlargement of cells already present or a combination of the two methods. Fruit size and fruit weight do not always keep pace with one another, and some fruits get less dense as they develop. In ripe apples as much as 25 per cent of the volume may actually consist of air spaces and most gardeners are already aware that many fruits contain well over 90 per cent of water in their tissues.

RIPENING

When a fruit reaches its full size it is said to be mature. From then on the processes of ripening take over and continue until the fruit is fully ripe, in the edible sense. If not eaten the fruit will continue into the over-ripe state, which is usually terminated by the invasion of micro-organisms that assist in its final decay. Ripening is not a period of gradual slowing down of the life processes of the fruit. Instead it is a period when new substances are being produced, insoluble reserves are being mobilized and all manner of complex changes are taking place. During ripening the tissues usually soften, due mainly to changes in the walls of the cells which form the fruit and a loosening of the pectic 'cement' which hold the cells together. Starches and fatty reserves are changed into sugars. New pigments develop and the green chlorophylls are broken down. The formation of some of these new pigments, for example the red pigments of some apples, is stimulated by light and that universal 'trigger' phytochrome is involved. New flavours also develop during ripening.

Many years ago it was demonstrated that ripening of some fruits—apples are again a good example—is accompanied by a dramatic rise in the rate of respiration. This is known as the *climacteric* and it eventually reaches a peak and then begins to decline as the fruit becomes senescent. In some fruits the climacteric is very rapid, in others more moderate in speed, whilst in some it does not occur at all. These types are exemplified by the banana, the apple and the orange respectively. Modern methods of chemical analysis have shown that the ripening of all these fruits is associated with the production of a gas called ethylene. In ripening bananas and apples large amounts are produced but in oranges the amount is hardly detectable. Most fruits will still experience the climacteric if left attached to the tree but ripening is often speeded up by picking the mature fruit.

The ripening of fruits proceeds fastest in the temperature range where growth is also most rapid. Ethylene appears to be a natural 'trigger' for ripening and the addition of traces of this gas to the atmosphere around fruit will speed up the ripening process. The old type of domestic gas supply often contained enough ethylene as an impurity to make it useful for speeding up ripening. When the fruit was placed in a closed container filled with gas, ripening could be speeded up by days or even weeks. Unfortunately for gardeners the 'natural' gas which is now used in many places contains virtually no ethylene and is useless for encouraging ripening. Natural gas does, however, have other virtues, which are mentioned in Chapter 9. Ripe apples can be used as a source of natural ethylene to promote the ripening of other fruit and apples can sometimes give trouble by causing bananas to ripen too quickly. A similar technique can be used with the apples as with the gas; enclose the slow-ripening fruit with the apple and wait! It should perhaps be mentioned that to the botanist tomatoes are just as much 'fruit' as are apples and pears.

FRUIT STORAGE

Not everybody wants to speed up ripening. Commercially there is great advantage in slowing it down. As might be expected low temperatures slow down the reactions which bring about ripening. The temperature must not be too low—certainly not below freezing point unless you want to go in for a poor quality frozen food. Pears are particularly susceptible to low temperature damage and their further ripening can be completely upset if they are chilled too much. The most satisfactory way of prolonging the life of a fruit over a long period of time is to store it in an atmosphere which is low in oxygen and high in carbon dioxide. This is the so-called 'gas-storage' of fruit and it aims mainly at reducing the respiration rate of the fruit. This, of course, slows the climacteric in those fruits which have one. The amateur who does not possess complex chambers for gas storage of fruit can usefully prolong the life of some fruits by storing them in sealed polythene bags, which leads to an automatic rise in the level of carbon dioxide and a fall in the level of oxygen inside the bag.

The danger of storage in closed bags is that the conditions may be particularly suitable for the growth of micro-organisms so that the bags must be regularly examined and any decaying or softening fruit removed immediately. Fruit should always be stored so that each fruit is not touching its neighbours. Wrapping each one up separately may help to stop mould spreading but it does have the disadvantage of having to unwrap each time you want to inspect the fruit. Only good, completely sound fruit will keep. Any damage will cause rapid decay which may then spread to previously sound fruit; so throwing a lot of fruit into a box is just asking for trouble. The preformed fruit trays, held on simple wooden racks, are very useful. Fallen fruit is almost invariably damaged and will not keep. The main danger with 'open' storage is that the fruit will dry up—particularly if you have it in a centrally heated house. In summary, prolonged fruit storage needs a cool, frost-free store which is airy so that dampness does not accumulate but not so dry that the fruit shrivels. If you can satisfy all those requirements you will be a lucky gardener. In conclusion it should be mentioned that some cultivars naturally store better than others so if you are buying new fruit trees this point is well worth considering.

Chapter Eight

Weeds, Pests and Diseases

When a gardener enters his garden he has a dual role to fulfil. On the one hand he is there to take care of his desirable plants, propagating, pruning and feeding them. On the other he has to fight a never-ending battle with the pests, diseases and weeds which threaten to invade and engulf his earthly paradise, The most successful gardeners in this battle are those who have learnt something of the ways of their enemies and who know how to recognize their friends when they seen them.

A garden, even a 'natural' or 'semi-wild' one, is not a fixed unchanging thing, and the more unnatural and exotic it is the more effort is needed to keep it in that state. For many years botanists have been studying plants as they grow naturally in the various parts of the world. This forms part of the subject known as ecology which, in essence, is the study of living organisms in relation to their surroundings. If plants are left to grow on their own, without any interference from man, it is generally found that in any one area there gradually develops what is known as the *'climax'* vegetation. Once this has been reached the species which are present will remain more or less constant for a long period of time. The particular type of vegetation found in a

152

'climax' depends mainly upon the climate and the type of soil. For a temperate country, such as Britain, climax vegetation would mean that much of the land would be covered by some form of forest, as it was, largely, when man first started to cultivate it. Since then he has been hard at work cutting down the trees and preventing new ones from growing in their place.

WEED BIOLOGY

Once a piece of ground has been cleared of its previous vegetation it is available for colonization by other plants. The trees do not reappear at once but are preceded by more rapidly growing plants, and only later when trees become dominant do many of the smaller plants die out as the amount of light getting through the leaf canopy is reduced. This change from one group of plants to another is known as a *succession*. The gardener is continually at work to prevent succession. He does not even desire the first natural colonizers. Instead he plants his own select species (see photographs on page 67). Natural succession results from competition between plants in any one area for the necessities of life, and it is by this act of competition that weeds affect the growth of garden plants. Weeds occupy space in the soil and air, and this is not then available to the gardener's plants. From the soil they take up water and nutrients, while their above-ground parts intercept light that might have been used for photosynthesis by the planted crop. Etiolation results as the plants struggle upwards. But these same effects can also be caused by planting the desirable plants too close together. Knowing the correct spacing for plants is part of the gardener's skill.

Plants which are noxious weeds to the gardener are often not of any great significance in the natural environment. This is because they are plants which colonize bare open ground rapidly but are not so successful in holding their own in an established mass of vegetation. In nature there is rarely any open ground

153

which can support plant life that is not immediately covered with plants. Among the first plants to appear on such ground are the annuals producing copious quantities of seeds, which are spread easily from one place to another by wind, animals and other agencies. Sometimes a plant's life-cycle is so short that several generations can occur in a single year. The number of seeds lying dormant in the soil goes on increasing until the growing plants of the species are ousted by other competing plants or by the arrival of the gardener. The seeds of most natural plants can remain viable in the soil for many years and each fresh cultivation by the gardener will bring more of them to the surface and into conditions favourable to germination. The old adage about 'one year's seeding seven years' weeding' is close enough to the truth for most gardeners not to need mathematical proof of it. Because of the enormous number of seeds produced by many weeds even a single plant can represent a long-lasting source of future trouble. The same rules apply to weeds as to garden plants. A small weak weed will produce fewer seeds than a large luxuriously growing one. To the gardener the most annoying thing is when the large weed is just over the fence in his neighbour's garden! Although laws do exist which require people to keep down the weeds in their gardens few gardeners feel inclined to invoke them, for obvious reasons.

To the practising gardener the perennial weeds are often the most troublesome. Given a few months' growth they can store up reserves to tide them over many onslaughts from the gardener. Their rate of spread by vegetative means is often phenomenal and a single seedling can soon colonize large areas with underground stems or roots. When cultivation is carried out even small pieces of the underground parts can regenerate strongly into whole plants (see opposite). Botanists even use some in their regeneration studies because of the ease with which they can be grown. Their penetration into the ground may be to a depth of several feet so that a light surface cultivation is rarely sufficient

154

to remove them. Although the perennial weeds give a lot of trouble by their strong vegetative growth most are also capable of setting seeds, to a greater or lesser extent, and this is often how an initial infection of the ground occurs.

WEED CONTROL

The control of a weed is usually assisted by a knowledge of its life-cycle: there may be one particular stage when it is most sus-

The noxious field bindweed can reproduce both vegetatively from small pieces and by seed

155

ceptible to damage by cultivation or weedkillers and attack at this stage is the most economical method of control. Another problem in weed control relates to the type of garden plants among which the weed is growing. If they happen to have the same habits of growth and the same type of life-cycle it will be much more difficult to eradicate the weeds and leave the garden plants undamaged.

Burning

Fire was probably the earliest method used by man to clear an area of its vegetation. In developed countries controlled burning is rarely an acceptable means of clearing the ground of tree cover but it still finds an application in scrub clearance where the danger of the fire getting out of hand is not so great. In the gardener's armoury fire is mainly used in two ways. Firstly, in the incinerator or bonfire he can destroy weed seeds and the parts of a plant which might give rise to vegetative spread. Unless a gardener is really confident of his ability to build a compost heap which will heat up sufficiently to kill such propagules it is far wiser for him to burn weedings which contain seeds or pieces of underground rhizomes or spreading roots. Many gardeners unwittingly spread weeds about the garden with poorly made compost. The second use of fire in the garden is to attack the weeds, where they grow, with the flame gun. There are many different types of gun: some clear a broad strip while others are fine enough to use on single weeds between the garden plants. There is a certain satisfaction in actually seeing the weeds shrivel up before one's very eyes! Any seeds will almost certainly be killed and as there is usually no need to clear up the burnt remains the nutrients are returned to the soil in the spot from which they came.

As with any gardening method of weed control there are dangers and limitations in using a flame-gun. The risk of catching the wrong thing alight is obvious. Weeds actually among the cultivated plants canot be tackled. Deep-rooted perennials or those with underground rhizomes will normally be rather difficult to kill and will be able to regenerate in the absence of competition from surface weeds. The depth of killing can be increased by using the flame gun on ground which has been dug over to expose the underground parts, and flaming of the soil surface will also kill the weed seeds on the surface. The depth of penetration of the heat from a flame gun is usually fairly limited, which is a disadvantage from the viewpoint of weed killing but an advantage in preventing damage to the soil structure by excessive heat. Many readers may have noticed the poor growth of plants in the very centre of an area previously used for a large bonfire. This is due to soil damage caused by the extreme heat. It should not in fact occur with a correctly used flame gun because the temperature needed to kill the weeds is less than that which damages the soil.

Even the gardener with a small garden can usefully possess a flame gun to cut down his labours, but it must be chosen to suit his particular requirements and used strictly in accordance with the instructions. Flame guns, of course, can be used for soil sterilization and are not limited in their usefulness to weed control. One final botanical point is that the dormancy of certain seeds is broken by heat and burnt soil therefore frequently develops weeds which have been stimulated into growth by the heat, where this has not been quite sufficient to kill the seeds.

Cultivation
Physical disturbance of the soil and weeds is probably still the most common method of weed control. Digging brings weed seeds to the surface where they can germinate; but with further digging the young seedlings can be buried where they will die and provide humus to the soil. By deeper digging the depth of good soil is increased and the chance of regeneration by the buried weeds is reduced. The sheer stamina needed as well as the

depth of top soil usually limit the depth of digging! There is little point in incorporating vegetable matter in the soil beyond the rooting depth of the plants you are going to grow, nor in trying to dig down to bring up subsoil which has obviously never even supported the growth of the weeds. If you are going to dig it should be done well before planting so that you have a chance of getting rid of the weeds before the garden plants go in. Sites for perennial plants need to be cleared of perennial weeds well in advance because later cultivation will be restricted to the zones between the plants. Obviously weeds should never be allowed to flower and set seed. Even if full weed eradication cannot be carried out seeding must be kept in check.

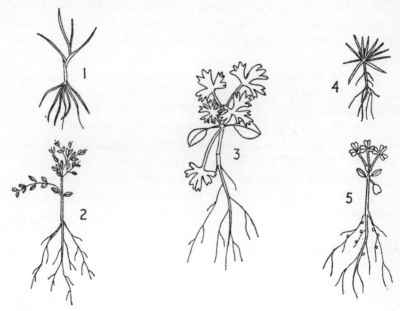

An important part of a gardener's training is learning to recognize weed seedlings: (1) annual meadow grass, (2) chickweed, (3) buttercup, (4) pearlwort, (5) trefoil, with root nodules

Once the surface of the ground has been cleared of weeds further digging merely serves to bring more seeds up to cause trouble and should be avoided if possible. Even such shallow cultivation as hoeing can also cause damage to surface rooting plants and the use of 'non-cultivation' methods is gaining some acceptance. In 'non-cultivation' the soil is disturbed as little as possible and the technique is obviously of particular use among woody perennials. Most gardeners would frown upon it both for visual reasons and because the formation of a hard surface layer can cause watering difficulties, so it is rarely suggested in gardening books. It does, however, link economically with the use of certain herbicides.

Competition

Weeds can sometimes be kept in check simply by competition from garden plants. This is an example of 'biological' control. If we bring in well-grown seedlings they will have a distinct advantage over young weed seedlings just developing in the soil; all the usual features of competition occur but this time to our advantage. Another method is to use ground-cover plants to fill in between shrubs and trees and to interplant one crop with another so that bare patches are not left between plants. Interplanting can be a little troublesome if it is not thought out properly because its restricts access and soil cultivation between the plants. But intelligent use of ground-cover plants can reduce the effort needed to keep a garden tidy and can positively improve its appearance.

It should, perhaps, be mentioned that competition is not always a question of plants needing the same water and minerals from the soil and sunlight and air from above ground. Certain plants produces substances that actively inhibit the growth of other species, but although many examples seem to exist as judged from visual appearances alone botanists have not yet been particularly successful in isolating substances which are exuded by one plant and can affect another in this way. That growth or germination can be adversely affected in this way has been proved beyond doubt, even if the exact substance involved is not known. Once again there is sense in taking the advice of reputable gardeners who have experience to draw upon. Remember that there are plenty of other reasons why one plant will do well and another species poorly when grown next to it, so try not to jump to conclusions concerning toxic interactions between plants.

Herbicides

The use of herbicides (weedkillers) in gardens has become extensive because of their effectiveness and convenience. The profesional grower has a much more extensive armoury of chemicals to choose from because he normally has the correct equipment to apply them safely and large areas of the same crop on which to use them. In the garden skill in herbicide use may be limited and the mixing together of several different species in a small area also presents problems. Garden herbicides are therefore those which are both safe to use and suitable for the particular applications of the amateur gardener. If you read about a substance used professionally and find you cannot obtain it, you may be sure that it is not available for very good reasons. The dangers of 'scrounging' a supply from a professional friend are too great to risk; even if you do not poison yourself you will not know its exact potency or suitability for garden plants and may well kill everything by mistake. The exact details of strength and method of application must always be adhered to when using herbicides. This is because the amount of the active substance used will determine which plants are damaged, how long any residual substance will remain active, and the cost of the treatment. It may surprise many gardeners to know that the effectiveness of a weedkiller depends to a marked degree on the way in which it is applied and it is not simply a question of having a certain strength of the active substance applied per unit area of ground. The formulation (the exact 'mixture' in the bottle) used and the method of application are of vital importance. So also is information concerning the weather conditions and timing of the treatment. Many 'new' garden weedkillers actually contain well-known herbicides in a new, more effective, formulation rather than completely new active substances. In fact the number of different active substances available to the gardener is very limited.

Herbicides can be classified in various ways according to their mode of application, their chemical nature, their biological effectiveness and so on. Total weedkillers are those which will kill all plants and were the first type to be used extensively. They can be used to clear rough ground and to keep paths, car parks and the like free from weeds. Most of them are very persistent and this is an advantage in many applications. It may be many months before ground treated with them can be used for garden plants so their gardening uses are rather limited. Sodium chlorate is one of the most extensively used total weedkillers but because it has a high fire risk all commercial preparations contain fire preventives as well. For many garden applications a selective action is required. 'Lawn sand' which contains a mixture of ammonium and ferrous (iron) sulphates is a simple example of a weedkiller which is selective mainly because of its mode of application. The broad-leaf weeds which form rosettes such as the dandelions and plaintains trap the granules of the sulphates close to their growing points so that a toxic concentration exists. The grass, on the other hand, does not trap the active ingredients in this way and is therefore exposed to a more dilute concentration. The nitrogen of the ammonium acts as a fertilizer to the grass and gives it a competitive advantage over the ailing weeds. The sand is largely an inert carrier to assist the spreading of the sulphates.

Resulting from work on the plant auxins there is a whole range of 'synthetic' auxins with herbicide activity. These are the true 'hormone' weedkillers such as 2,4-D, 2,4,5-T, MCPA and so on. They were the first truly selective weedkillers with a wide application in weed control and each has its own particular uses. In most cases the 'hormone' weedkillers are selective because of their greater absorption by some plants than by others, and again there is a greater tendency for them to be effective against broad-leaved weeds. They have the advantage of being capable of translocation (movement) within the weeds so that they can penetrate down into the underground parts of perennials. For this use a weak application is advised which will not kill the above-ground parts too quickly before they have had time to absorb enough to accumulate in toxic concentrations in the underground parts. The chemicals mentioned so far are toxic in their own right. A development from them are substances such as 2,4-DB and MCPB which are not themselves toxic but are converted by some plants into the related toxic compounds, in this case 2,4-D and MCPA, so the weeds literally kill themselves. Some plants are resistant to the hormone weedkillers and scientists have therefore developed a whole range of different weedkillers which exploit other aspects of a weed's physiology.

Commercial firms which develop new herbicides usually examine several thousand new substances each year. From this vast number perhaps one eventually reaches commercial use. This is because it must not only kill the weeds for which it is required but must also be safe to use. Safety in use covers a whole range of requirements. Firstly it must not present a toxic hazard to man and animals. Secondly it must not be toxic to the crop in which it is to be used. Thirdly, unless a persistent weedkiller is desired it must break down rapidly so that it does not accumulate in the soil or in animals which eat the sprayed plants. Fourthly, it must not upset the balance of organisms in the soil because this could destroy soil fertility. We could go on extending the list. Because of the high cost of development few commercial firms can afford to develop a totally new active compound specifically for garden use. The new substances are mostly designed to meet a specific requirement in the protection of an important economic crop. Subsequently specific formulations of that particular substance may be developed to meet the needs of the amateur gardener.

Weeds, of course, are not completely fixed in their characteristics, and it may ocasionally be found that a substance which was initially effective gradually becomes less able to control a given species because new strains of the weed have developed which are more resistant to the herbicide. Even when a herbicide is fully effective against its target weeds the problems are not over. The space formerly occupied by the weed may become filled with another species which is not affected by the herbicide and grows rapidly now that its competitor is removed. So the battle is never completely won. It is probably an advantage that weedkillers are not quite as effective as some might hope. Many insects depend upon weed species for their survival and the complete elimination of any species of plant from all its *habitats* (the places in which it occurs) might well have undesirable effects— certainly beekepers find 'weeds' an advantage to their craft.

PEST BIOLOGY

So far we have considered plants in isolation from other organisms but, in fact, they always form part of a larger unit called the *ecosystem*. The green plants, be they weed or cultivated, are the main *primary producers* because they can create new organic materials out of simple inorganic substances like carbon dioxide, water and minerals. All animals ultimately rely upon plants to obtain their energy ('food'). Some feed directly on plants or their remains while others feed on these animals, and so on. This gives rise to the concept of a *food chain* which is simply a list of the animals present in the order in which they eat each other right down to the green plant at the end. The alternative concept of a *food pyramid* is better because the passage of 'food' energy from one organism to another is not 100 per cent efficient. It therefore takes a very large number of green plants to support a smaller number of direct plant-feeding animals, and these, in their turn, support a smaller number still of animals which eat them, and so on. (In the present context 'animal' is used in the zoological sense and includes insects and other invertebrates as well as birds, reptiles, mammals and all the rest.) Even closer to the truth is the idea of a *food web* because all the living organisms in an ecosystem interact with one another, either directly or indirectly.

Until quite recently scientists could do little more than merely list the species in a given habitat. More detailed research is now revealing the way in which the numbers of each of the different orgnaisms can fluctuate. In an ideal 'natural' environment, which to all intents and purposes now hardly exists anywhere in the world, the number of individuals of a species usually shows cycles of increase which are then stopped by lack of food, a disease, some predator, or a change in climate. Then follows a decline in numbers to a much lower level from which an increase may again occur. The time has long since passed when man's influence was no more than that of other living creatures. Today, his skills enable him to create very artificial environments, such as a whole field of the same plant, and naturally this often leads to an enormous increase in the numbers of some creature which lives on that cultivated plant. Man's own two hands are not sufficient to control such plagues and he cannot afford to write off the crop. So it is necessary for him to find other methods to kill insects, mites, and so on.

PEST CONTROL

The problems are exactly the same as those of weed control. The early pesticides were not very selective and killed off both harmful and useful insects. Selectivity can be obtained, to a certain extent, by using the pesticide in a particular way and at a particular time. Problems caused by the development of pesticide resistant strains seem particularly severe among the invertebrate pests, such as the insects.

There are a few simple techniques which can be used to help in the battle against pests that do not involve spraying with pesticides. These make use of some particular characteristic of the pest itself. Female winter moths, which do not have wings, crawl up from the ground to lay their eggs on apple trees and if a sticky band is tied round the tree trunk the insects are trapped on this. Earwigs like to crawl up and into dark crevices and will congregate in upturned cartons held on sticks.

As with herbicides, so the search goes on for more selective pesticides. Most of us are prepared to put up with a few maggots in our home-grown apples but few customers will accept them in produce bought in the shops. This has important consequences for the future of biological control of insects and the like. Biological control exploits some other creature from the food web which lives upon the pest, and this beneficial organism, if it kills off all the pests, must also itself die out because it will have no food. By its very nature biological control can never be 100 per cent effective. Another problem is that we usually do not know enough about the organisms involved to be sure of the result when we start using methods of biological control. True, it may work well when tried out in an experimental area but all sorts of things can go wrong when it is used on a wide scale. One of the most interesting aspects of pest control at the present time is the co-ordinated use of biological control with limited amounts of selected pesticides. This may well give the best of both worlds but it does, of course, need much greater skill and knowledge than just giving the plants a spray over with a pesticide that will kill everything. Another method of biological control which is being used is to infect some of the pest organisms with a virus or bacterial disease which will spread from one individual to another and eventually kill off most of them.

One of the most serious problems with the early pesticides

such as DDT is their extreme persistence. At first this seemed an advantage because the active ingredient does not break down and lose its effectiveness too quickly. It is now obvious that they are too persistent because they gradually accumulate in the bodies of animals which eat the pests and may eventually kill these animals as well. Newer chemicals are being developed which break down to harmless substances once they have killed the target pest. Another problem with pesticides for use against animals arises because man is closer to animals in his physiology than he is to plants. It is therefore not so easy to find insecticides harmless to man. Those whose business is pest control are always on the lookout for new and safer methods. It is possible, for instance, to produce substances which attract certain insects, and in this way they can be lured to their death. Another method is to release large numbers of sterilized insects which will compete with the normal wild ones in mating but will be unable to produce any viable offspring. Faster progress might be made if those who are at present so vociferous about the dangers of pesticides were to spend their time instead in trying to develop better ways of pest control.

There are a few cultural practices which will help the gardener in his fight against pests. Removing and burning infested shoots before the pests can spread too widely is of great benefit, and so is the keeping down of weeds which might shelter pests. Pests, of course, are not restricted to the aerial parts of plants and roots and shoots can be attacked from an infected soil, which may need complete sterilization. The only alternative to this may be to grow plants which are not susceptible to the pest: many such plants do exist because fortunately pests tend to prefer only certain species of cultivars of plants. A gardener relying on natural recovery of an infected soil may have to wait many years before it has improved sufficiently for the growing once more of susceptible crops. A form of biological control can sometimes be obtained by planting a crop which the pest will not attack or which produces a toxic exudate. *Tagetes* has been used in this way to improve soil infested with eelworms. Another aid to control is to encourage natural predators, such as birds, in the garden. It is a question, however, of which birds, which crop and which invertebrate pests you have.

DISEASE BIOLOGY

The so-called 'diseases' caused by deficiencies of nutrient elements and others caused by toxic substances have already been mentioned elsewhere in the book. Diseases in the more precise meaning of the word are caused by fungi, bacteria or viruses. The key to their control is to understand the way in which each multiplies and spreads. Nearly all fungi produce spores at some stage in their life cycle and by this means spread from plant to plant. Although some fungi (the *obligate* parasites) can only grow on living host plants others (the *facultative* parasites) can also grow on dead material and may even normally do so. There are also many fungi, known as *saprophytes,* which live entirely on dead material and are not known to cause any plant disease. A fungus that only grows on a living plant is not necessarily the more serious threat because the facultative parasites can live on dead plant remains lying about the garden.

For every garden plant there are over a hundred different fungi that can attack it and most disease micro-organisms can grow on more than one plant. Fortunately the outlook is not so black as it might seem because many fungi produce only minor symptoms and in practice it is found that only a few diseases really cause trouble in the garden. One consequence of a micro-organism having several hosts is, however, that perennial and wild plants can carry the disease over from one season to another. A few micro-organisms actually need a period of growth on two different plants to complete their life cycle and these can be controlled by eliminating one of the two host plants.

DISEASE CONTROL

Many plant diseases which give trouble in the garden can be dealt with simply by examination of the affected plant and consultation of the relevant books on plant pathology. Symptoms need to be noted first. Is it a rot, canker, leaf spot and so on? Then the organs affected must be noted. Finally the species of plant infected must be known. By this means the amateur can arrive at a reasonably satisfactory diagnosis, though the professional botanist might wish to go into further detail before he felt certain. Some diseases like apple scab can be controlled satisfactorily by the recommended methods at very little cost. But with others the cost of control may be more than the plants are worth and the gardener may decide simply to dispense with those plants altogether. There are even some, such as dutch elm disease, for which no satisfactory cure is at present available and only cultural techniques and general cleanliness can keep them within limits.

One of the best ways to avoid damage by disease is to breed resistant cultivars of plants and work of this kind has been going on for decades. However, just as a plant's resistance can be increased by selective breeding, so can the disease organism change its mode of attack. Where detailed studies have been made it is usually found that there are several strains of any one disease, each able to infect various cultivars, and new strains can arise almost as quickly as new cultivars are bred.

Resistance to disease is a genetic characteristic which is passed on to the offspring and can be transferred from one cultivar to another by selective breeding. The resistant plant may have some structural feature which prevents the disease developing, such as a more effective protective layer over its surface. Changes in the plant's physiology can also cause resistance. Cork may form around sites of infection and isolate them or abscission layers may form which cause the infected part to fall off. Limited necrosis

168

(death of the cells) can also act as a barrier. Some plants naturally produce substances which inhibit the growth of disease organisms or produce such substances as a reaction against infection. The susceptible plants are generally those which do not react sufficiently rapidly or strongly, or react in an ineffective way, towards the invading organism. The diseases, of course, have their own armoury against the plants and can, depending on the organism, produce enzymes to soften and dissolve the tissues, toxins to kill them, hormones to upset their normal growth and gummy substances to upset the flow of water and nutrients.

The conditions under which a plant is growing can markedly affect its resistance to disease. The climate often influences the incidence of diseases, as temperature and moisture play a role in infection. The pH value of the soil is often an important factor. Clubroot of brassicas, for example, is favoured by acid soils but potato scab is less serious if the soil is acid. Sappy growth due to excessive nitrogen fertilizers invites many diseases, while other nutrients, such as calcium, appear to improve resistance; this does not mean, however, that huge applications of calcium or any other beneficial element will control diseases.

Good hygiene in the garden is an essential for encouraging healthy plant growth. This means clearing up all dead and decaying leaves, twigs and plants. Any which are obviously infected should be burnt and only disease-free material incorporated into the compost heap (correctly made). Any badly infected parts should be cut off or removed from growing plants before the disease has time to spread. Rotting fruit should never be left around on the ground. Any large wounds or pruning cuts must be covered with protective paint, preferably as soon as they are made. Badly diseased leaves should be taken off at once and the autumn clean-up made before diseased leaves have a chance to rot into the ground. But as even all these practices will not eliminate every disease it is fortunate that there are also fungicides available, ranging from simple lime-sulphur mixtures to the

169

L

newer types designed for specific disease problems. (See photographs on page 135.)

Viruses are so simple in their structure that they can only develop in living cells because they need to use some of the 'machinery' of the host cell in order to multiply. They are spread by contact between plants and by small invertebrates like aphids which act as vectors (carriers) from one plant to another. Some viruses have little or no effect on the health of the plant but others cause very serious diseases. Every part of a virus-infected plant should be removed and burnt as soon as the diagnosis is certain or even before; weak plants that do not seem to be suffering from any obvious shortage, nutritional or environmental, are best removed as they will never prove satisfactory. Because of their intimate association with living host cells viruses are difficult

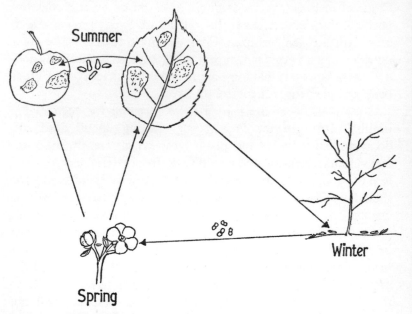

The methods by which apple scab can spread and overwinter in leaves

170

to control. It is possible to grow resistant cultivars in some cases. With a few plants meristem culture can be used to obtain virus-free stocks, but obviously these must then be protected against reinfection.

Many important plant diseases, such as fire blight of pears, soft rots of vegetables and canker of stone fruit trees are caused by bacteria. Bacteria increase in number simply by splitting into two new cells and this can be very rapid under conditions favourable to the bacteria. Some, like the bacteria causing fire blight, re-produce mainly in the host plants but other bacteria, particularly those causing soft rots, increase in numbers in the soil as well. Although bacteria are not so intimately bound up with the host plants cells as are the viruses, bacterial diseases often prove equally difficult to control. They spread from plant to plant mainly with the assistance of vectors such as insects, animals and man himself. Rain splashes and moving water can also spread some bacterial diseases. If the gardener has to handle infected plant material or rotting fruit he should wash his hands and tools afterwards to prevent bacteria being spread to any sound plants. Generally bacterial diseases do not respond to control by chemical sprays and a combination of clean cultural methods and the plant-ing of resistant cultivars is usually recommended. Where sprays are used copper containing ones seem to be among the most use-ful and there are also many experiments being carried out to investigate the suitability of antibiotics for the control of bacterial diseases of plants.

Not all micro-organisms are harmful to plants. The continuing fertility of a soil is dependent upon the activities of 'benefical' organisms which break down dead organic remains. Others have already been mentioned which can 'fix' nitrogen from the air. There are also many that can kill animals which attack plants. Nematodes (eelworms) in the soil can be killed by the various nematophagons fungi, some of which have very complex trapping mechanisms to catch the nematodes, and as previously mentioned

171

many insect pests are subject to diseases and these can be used to effect a biological control of the insect. It is for these reasons that a 'total' persistent fungicide would be as harmful as a 'total' persistent insecticide. Fortunately resistance to fungicides does not appear to be too big a problem at the present time but it could come if they are used indiscriminately. Good gardening practices plus the intelligent use of chemical control methods are the keys to success against all weeds, pests and diseases.

Chapter Nine

Greenhouses, Frames and Cloches

The range of plants which a gardener can grow in the open air and the season of the year when they will grow are limited mainly by the climatic conditions of the area in which he lives. It is therefore obvious that the keen gardener will seek ways of changing the climate so that his plants will grow earlier or later in the year and so that he can grow plants from countries with climates different from his own. It is not possible to vary the conditions appreciably in the open air, at any rate on a large scale, so various types of construction are necessary to provide a micro-environment and to allow some control to be exercised over it. The important effects of soil, temperature, water, light and air upon plant growth have been mentioned in earlier chapters. Of these factors, temperature and light are the two which mostly restrict plant growth. If the gardener records or looks up the average temperature and hours of daylight throughout the year he can construct a climatic 'orbit' for these two important factors (see diagram over). The gardener will only be able to grow successfully those plants whose physiology is such that they can develop to the required stage within the limits of the climatic orbit. Some form of controlled environment will be necessary if he wishes to grow plants whose requirements extend outside those limits.

THE THEORY OF ENVIRONMENTAL CONTROL

Before the details of environmental control are considered some important general principles must be explained. The way in which a plant develops is dependent upon the genetical make-up of the plant (the species and cultivar characteristics) and the conditions under which it is growing. The separate environmental

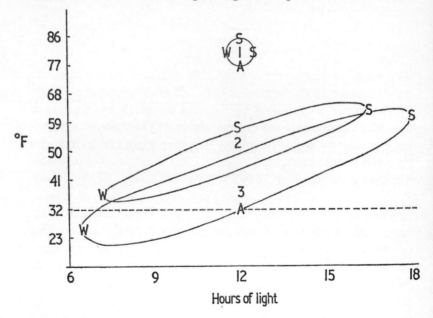

A simple climatic orbit for: (1) a tropical region, (2) a temperate region with mild winters, and (3) a region with severe winters. Monthly averages are used so that even in (2) there could be some frosts

174

factors do not, however, act in isolation from one another but interact in a very complex way. For decades botanists have thought of photosynthesis in terms of the concept of limiting factors which simply means that the rate of photosynthesis is limited by whichever requirement is in short supply. The amount of light, and of CO_2, the temperature and so on all affect the rate of photosynthesis but if one of these factors is increased while the others are kept constant the rate of photosynthesis at first increases more or less in step with this factor and then slows down and eventually stops, even though the varying factor is still increasing in its amount. This is because one of the other factors is now limiting photosynthesis and a further increase in the rate will only occur if the 'limiting' factor is increased. This concept can be extended to cover every aspect of plant development— there is always some factor which is preventing faster growth, causing etiolation or preventing the desired processes of flowering or fruiting. The importance of all this to the gardener who wants to control the environment of a plant is simply this—it is no use increasing the amount of something which the plant already has enough of. Raising the temperature in a greenhouse will not make the plants grow faster if they are short of water or if the light is not strong enough. In most cases plants can cope with variations in the outside environment because they have evolved to suit the natural way in which one factor varies with another. Low light levels in winter are matched by lower temperatures. In the greenhouse low light levels plus high temperatures may spell disaster.

The building-up processes of photosynthesis and the breaking-down processes of respiration have already been mentioned. For genuine growth to occur rate of photosynthesis must be greater than that of respiration. What is known as the *compensation point* occurs when the building-up reactions of one balance the breaking-down by the other. Below this the plant's reserves are being used up. The exact value of the compensation point varies according to conditions inside and outside the plant and can only be stated in terms of the amount of light, the temperature, the supply of carbon dioxide, the species of plant and so on. At low levels of light an increase in temperature speeds respiration more than it does photosynthesis. For example, 195 lux is the compensation point light intensity at 15°C (59°F) for *Pelargonium zonule* growing in air. If the temperature is raised to 25°C (77°F) the light intensity must be raised to 490 lux merely to keep the plant at the compensation point! Commercial growers who must reckon the cost of production are very concerned with keeping plants above the compensation point. If a plant is kept below the compensation point for long periods not only will it not grow but also the leaves already formed may turn yellow and become senescent. In lettuce it has been found that increasing the concentration of carbon dioxide in the air lowers the compensation point so that the plants can be kept growing in winter at higher temperatures without the need for supplementary light. All this sounds complicated and indeed it is. Those who want to turn winter into summer simply by buying a greenhouse will meet plenty of snags. With many plants, particularly from temperate climates, prolonged exposure to temperatures above 25°C, even in the presence of adequate light, results in a reduction in rate of photosynthesis. The higher the temperature the more rapidly does the decline become apparent. The rate of respiration usually only declines at rather higher temperatures. In natural outdoor conditions and using the plants suited to the particular climate these high temperatures will rarely be reached; but in a greenhouse, with inadequate ventilation, temperatures rise dangerously on very sunny days.

THE STRUCTURES THEMSELVES

There is no such thing as the 'all-purpose' greenhouse in which every conceivable type of plant can be grown at the same time.

If the gardener is setting out to control plant growth he must first decide which plants he intends to grow and then obtain the greenhouse to suit them. If he already has a greenhouse he must find the plants which will survive under the particular conditions which he can afford, both financially and in terms of the time taken to look after the plants.

It is technically possible to provide the ideal conditions for almost any single species of plant. The botanist can do this in his experiments by using a plant-growth chamber, of which there are many different types available. But the cost of buying and maintaining full control over a shelf area of even a few square yards is much more than that of the average man's own house. Larger areas or cheaper installations invariably mean less control over the conditions in which the plants are growing. To really understand a plant's physiology a whole series of different growth chambers are required, as provided by the 'phytotron', which is a series of accurately controlled greenhouses, or even a collection of small growth chambers grouped together in some particular place. The few really large installations such as those in America and France are world famous. Almost all reputable botanical work on the factors which affect plant growth is carried out in some form of controlled environment set-up. There are so many possible causes of poor growth that even if a certain factor cannot be controlled within pre-set limits by the botanist the extent of its variation must be monitored and recorded to help in the interpretation of the experiment. Eventually out of all this detailed scientific work come the recommendations for the best cultural conditions for a particular plant and it is then the gardener's task to try and meet the requirements. At the present time, of course, very few plants have been studied in sufficient detail, and gardeners must frequently regret that some of the plants found suitable for botanical studies are strains of wild plants rather than 'valuable' garden plants! But back to the ordinary gardener and his greenhouse.

The simplest types of structure, the various cloches, seek only to extend the seasonal limits for a plant by a few weeks. This is done by keeping the temperature in the micro-environment a few degrees above that outside. This is possible because of the particular properties of the materials of which they are constructed. Light in the visible and far-red regions of the spectrum must enter readily. This happens if clear glass or plastic material is used, but any 'clouding' of the material due to the presence of dirt, moisture or the nature of the material itself will cut down the amount of light getting through and it is vital to the health of the plants that as much light as possible does get through. Apart from the 'visible' light which is necessary for the plant's normal development some of the sunlight also consists of ultraviolet and infra-red rays. The ultra-violet is largely stopped by glass but as it is of no proved value to plants and may, indeed, actually inhibit plant growth this is no loss. The infra-red light can penetrate the glass and inside the cloche some of its energy becomes converted into heat. The result is a rise in temperature inside, with the slightly insulating properties of the glass preventing the too rapid loss of this heat. Small, sump-type paraffin heaters or soil heating cables can be used to supplement the natural heat from the sun. Inside the cloches humidity will rise as moisture evaporates but is not swept away by dryer air. Occasional opening of the closed ends of a series of cloches may be necessary to prevent damage by mould growth encouraged by damp and stagnant conditions. A certain amount of air flow may also help by preventing the carbon dioxide content of the air falling to a limiting level during sunny periods. Usually the ground inside the cloches will not need watering because enough water will be drawn, by capillary forces, from the damp soil outside the cloche area. Naturally only hardy plants can be grown in unheated cloches over the winter months when hard frosts occur but cloches are one of the cheapest and least time-consuming ways of extending the growing season a little and also

178

possess a great advantage in that they can be moved to another area when their present use is ended.

Those who want to help their plants even more and to extend still further the scope of their activities will almost certainly find a garden frame more economic than a greenhouse. The more substantial construction of the frame scores over the cloche in terms of durability and usually also in the insulating properties of its materials, although this does depend upon the type used. Located in a fixed position, the frame can take thermostatically controlled soil heating cables—a most attractive proposition. But whatever the form of heating adopted it will be far more economical than the heating of a greenhouse with its much larger volume yet usually no greater growing area. Adequate ventilation is also far easier to obtain in a frame than within a greenhouse. For hardening off plants from a greenhouse, bringing on a few trays of half-hardy seedlings, and for a multitude of other uses a garden frame is almost an essential to the keen gardener.

More sophisticated control of a plant's environment is possible with a greenhouse but this does not mean that all greenhouses represent the 'Garden of Eden' for plants. Greenhouses can be classified according to the temperature maintained in them. There is the unheated greenhouse in which only hardy plants resistant to frost damage can be grown in the winter, and in which, in the spring and autumn, an extension of the outdoor growing season is possible. In the summer the temperature fluctuations inside are likely to be so violent and the upper limits so high that few plants will survive unless great care is taken to provide adequate ventilation and some form of shading. This overheating problem also occurs in all other types of greenhouse. The next stage up is a frost-free house at something just over 1.7°C (35°F) in the winter. The lower limits of a 'cool' house usually range somewhere between 5.5 and 7.2°C (42-45°F), for the 'warm' house between 12.7 and 15.5°C (55-60°F) and the hot house 21.1—23.8°C (70-75°F). Few species of plants would be really happy in more than one of these different types. Classification can also be according to the actual construction of the building: into plant houses, forcing houses, dutch-light types, lean-to and so on.

HEATING

Apart from the trapping of the energy in the sun's rays, a greenhouse can be artificially heated by a number of different methods. The final choice depends on several factors, some of them not directly relevant to the plants themselves. If the gardener is not on the spot throughout the day some form of automatic control is an essential. Without this the apparatus may go on heating longer than is beneficial to the plants or the gardener's pocket, or the heat may not come on when it is needed. So before choosing a heating scheme it is essential to examine one's own habits as well as that of the plants! It is not my intention to discuss the details of each system. From the plant's viewpoint the temperature measured on the thermometer is not the one of importance to it. Intense local heat in one part of a greenhouse and cold spots elsewhere will not lead to even growth. Alternatively, a constant blast of hot dry air onto the leaves will cause transpiration to be so rapid that damage is likely to occur due to water shortage. Those who can remember a little school physics will recall that warm air rises and cold air sinks down towards the ground if left undisturbed and to prevent this some air flow, which is usually provided by ventilators or their equivalent, is essential in a greenhouse. Other reasons for admitting air from outside into the greenhouse are to replace the carbon dioxide used up by the plants and prevent the air becoming too humid. The actual position of the ventilators and the details of their control can markedly affect the amount of heat lost from the greenhouse, and this matter should certainly receive some attention before purchase as it can rarely be remedied afterwards.

Although having a thermostatic heating system will prevent

the temperature dropping below some pre-set limit it will not stop the temperature rising to excessive levels if there is a sunny spell, and another function of the ventilating system is to keep this rise within limits. With simple, hand-operated, ventilators the gardener must make the prediction. Thermostatic controls can also take over this task reliably—proper ventilating fans usually prove more satisfactory than simple open-shut ventilators which rely on natural air flow. From the viewpoint of the plant it is usually suggested that a complete change of air every minute is near the ideal but the cost of heating the air at this rate of flow may not be ideal for the gardener. Nevertheless it does give an indication of the importance of having a really adequate air flow.

All forms of heating which rely on burning a fuel within the greenhouse itself obviously consume the oxygen of the air and, ideally, produce carbon dioxide which the plants can use as the raw material for photosynthesis. All burners need good ventilation to the flame when in use because smoky, incomplete combustion can cause disaster due to highly toxic substances liberated into the air. These can be prevented from reaching the plants by ducting the fumes out of the house, though a little heat may also be lost. The advent of 'natural' gas will prove a boon to those who like gas heating because it is particularly low in toxic substances and can even be used without a flue. Some plants, of course, are much more sensitive to air pollution than others so many gardeners get away with bad systems if they are content to grow only a limited range of plants. The excessive temperature rises of summer due to the sun's rays penetrating the greenhouse and the consequent risks of overheating can be guarded against by making the glass opaque with a 'white-wash'. This of course is rather a disadvantage on cloudy summer days and must certainly be removed for the winter when the plants will need all the light they can get. Blinds of various types can be obtained which will do the same job and can even be fitted with thermostatic or photometric controls which will roll them up or down as conditions change.

WATERING

Water is often a problem in the greenhouse. If there is not enough the plants will wilt or even die of desiccation. Although wilting is often reversible and the plants recover turgidity quite quickly after watering it has been shown that their physiological processes, particularly with respect to the gaseous exchanges of photosynthesis, may take several days to return to normal. Remembering the relationship of relative humidity to temperature it is obvious that on a sunny day the air may become excessively dry inside even if water condenses out at night when the temperature falls. For the gardener with little time, or a poor memory, some form of 'automatic' watering is almost essentiial. Several systems are available, each with its own advantages and difficulties, to supply water to the roots, but in hot-houses some form of atmospheric damping is required as well. In the cooler greenhouse the problem may well be one of getting rid of the excess moisture which condenses everywhere and encourages the spread of all sorts of harmful micro-organisms.

The question often arises regarding the best type of water to use for pot-grown plants. Provided it has not run off a surface recently treated with a preservative or into a container which liberates toxic ingredients into the water there is little doubt that rain water is the best thing available. It contains little calcium, which, with prolonged use, can build up to harmful levels quite quickly in a small volume of soil. On the other hand it does contain some of the nutrient elements which plants require, its composition varying according to the place from which it is collected. Under trees it will contain greater amounts of nutrients leached from the leaves as it falls; near the sea it will contain more sodium and so on. Since rainwater butts are often full of decaying leaves this also helps to swell the nutrient content. This is not to say that a dirty water butt is to be encouraged! This can lead to trouble and it is normally much better to cover the butt properly and supply your own nutrients by way of organic or inorganic fertilizers than to rely on chance. Depending on the process used softened water may be better than tap water. Soluble water-softening compounds are of no use to the gardener as they will merely deposit again when the water evaporates from the soil or plant and therefore make the water worse than before.

Properly de-ionised water and genuine distilled water can be used on plants but it must be remembered that they contain no mineral nutrients and if used excessively, so that the water frequently drains out of the soil in the pots, they will leach out the soluble minerals as they go. Rather greater attention therefore has to be paid to mineral nutrition when using these than when using rain water. If you must use tap water direct watch out for the build-up of calcium if the water is hard. Some plants, of course, will not stand hard tap water at all. The pH value of hard tap water is often around 8.0 and that of distilled water sometimes around pH5.0. The distilled water has no buffering capacity and adjusts rapidly to the normal pH of the soil but hard tap water will tend to make the soil more alkaline, with all its attendant problems. It is possible to make up very dilute nutrient solutions to use when watering but the amount of nutrients must not be made too large or even they may build up to toxic levels in the soil.

To encourage the rooting of cuttings a very moist atmosphere is often needed to prevent them drying up before they get established. Mist propagation, as its name implies, uses a fine spray of water over the cuttings but the spray is not continuous since this would also lead to drainage problems with the excess water, not to mention the actual waste of valuable water. When the humidity falls below a certain level the spray is switched on again until the moisture content of the air has risen. The gardener could guess the time, if he happened to be around, but commercially available systems have various types of sensor which do the job automatically—and more reliably.

LIGHTING

So far we have discussed only relatively simple modifications to the environment within a greenhouse. With artificial lighting the gardener's problems really begin because he has to consider not only the amount of light produced but also the spectral composition of the light. In previous chapters the effects of light of different wavelengths were mentioned. Man's eyes are particularly sensitive to some wavelengths and it is not surprising that manufacturers of lighting equipment normally do all their calculations in terms of the way in which we humans see light. Since the growth of plants is affected in a complex way by many different wavelengths of light the amount which a man sees is often no guide to its effect on growing plants. If we just used sunlight there would be no problem because the sun's energy output stays at a fixed relative level for each of the different wavelengths and plants have developed on the earth in accordance with the energy distribution of the solar spectrum. Artificial lights may look perfectly good and bright to us but can cause abnormal growth of plants because they have too little or too much of the total light energy in certain regions of the spectrum.

For photosynthesis the wavelengths corresponding to red and blue light are usually most effective although in some plants red light appears rather more effective than blue. For phytochrome-controlled reactions the role of red and infra-red light has already been discussed (see page 137). Other plant responses such as phototropism are more promoted by blue light, and so on. In practical terms this means that if we are going to use artificial lights to assist our plants to grow we should be concerned not only with how bright they are from the human viewpoint but also with the spectrum of light which they produce (see below). Although there are only three common types of lamp which the amateur is likely to meet they can each come in a wide variety of shapes, sizes and special forms. Ordinary electric light bulbs have a tungsten filament in them. Only a small percentage of the electrical energy put through the filament appears as visible light and a high proportion of red and infra-red light is emitted. This means that responses controlled by phytochrome can be upset and abnormal growth usually results if filament lamps are used as the exclusive source of light. But for use merely in extending daylength the distorting effects on growth are minimal and this form of lighting has the advantages of not costing very much to

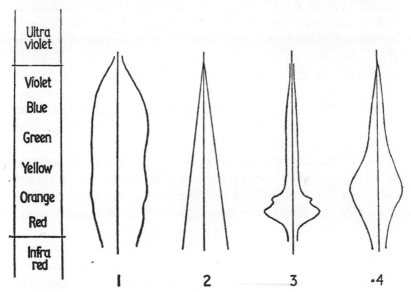

A very simplified comparison of the energy distribution for (1) sunlight, (2) an ordinary electric light bulb, (3) a mercury-fluorescent lamp, and (4) an ordinary fluorescent tube. The differences can have a marked effect on plant growth

185

M

instal and being easy to obtain. All electrical work in a greenhouse or frame must, of course, be waterproof. Fittings and wire intended for use in a house can be very dangerous if used in a greenhouse where moisture from condensation or watering can get to them. The gardener should remember that filament lamps give out so much heat that this can give rise to problems of overheating if they are placed too close to plants or in a confined area. It is of course possible to use an electric lamp as a heater for a propagating frame, but because of the danger from moisture the gardener is better advised to purchase a short length of soil-heating cable rather than risk his life in this way for the sake of a little money.

Although a simple mercury-vapour lamp gives out most of its visible light in the blue-green region, with an associated ultra-violet emission, the type of light emitted from the different specialised mercury vapour lamps can vary considerably. Some types are very suitable for plant growth, having a reasonable spread of energy over the required parts of the spectrum. Although special control units are neded for mercury-vapour lamps this has not proved a disincentive to installing them and they are used extensively by commercial growers, which is a clear indication of the lamps' value.

More familiar to most gardeners are the fluorescent tube lights. A variety of different tubes are available for use in the fittings and some are manufactured specifically for horticultural use. For their size the fluorescent lamps give out comparatively small amounts of light but they can be conveniently fixed in 'banks' close together because they also give out less heat. Control gear is required with fluorescent lamps, too, but many gardeners probably opt for this form of lighting more because it is 'the devil they know' than because they have seriously weighed its relative merits against those of the mercury-discharge types. The major disadvantage of fluorescent lighting is the great size and weight of a set-up sufficient to give high light intensities, but the spectral output from most types of fluorescent tube is quite suitable for nearly all garden lighting applications.

There are, of course, several other types of lamp, sodium and neon for example, which could be used for lighting, but these have found little favour by comparison with the three types already mentioned. It is perhaps worth mentioning in connection with these matters that although photographic light meters measure 'light' they are of no value in comparing different types of lamp for greenhouse use. They are designed to measure the light which affects photographic emulsions, not plants. The scales on them are also related to photographic uses. The measurement of light in a way which means anything in terms of plant growth is a very difficult matter. But the gardener should not be put off by the complexity of the subject; supplementary lighting has a valuable place in 'controlling' plant growth.

The control of daylength has two aspects: one, with which we have dealt, is to make the days longer with artificial lights; the other is to make the 'days' shorter by using blackouts. These are simple covers of blackout material which can be pulled out over the plants at the required time. Blackouts could be worked, as lights usually are, by time switches, but this usually necessitates a rather elaborate set-up. For one or two plants a cardboard box or light-tight cloth 'cap' can be used to shorten daylength.

Chapter Ten

Breeding New Plants

Man has been selecting plants for thousands of years. The history of several of our important food crops goes back to the dawn of civilization. At first this selection consisted of finding out which wild species were fit to eat or had some other desirable properties. Some of these species proved amenable to cultivation and then began the long process of selecting the best plants from within a species.

The concept of a species is fundamental to the study of living organisms but it is difficult to define exactly what a species is. At one time each different sort of plant was thought to have arisen by an act of 'special creation' so that each was uniquely different from all the rest. With the subsequent development of evolutionary ideas concerning the origin of the species some blurring of the lines separating different organisms occurred. In the early days of botany detailed descriptions and drawing of plants were made as a help in recognizing plants that the ordinary person knew by some common name. The problem with common names then, just as today, was that one plant might have several different names and the same name might be used for several different plants.

188

PLANT NAMES

It was not until the eighteenth century that the system of binomials ('two-word names') was developed, primarily by Linnaeus in his *Species Plantarum*. Plants which showed similarities were grouped together as a 'genus' and given a generic name. This is followed by the plant's own specific name which tells us exactly which plant we are dealing with. The generic name is always printed with an initial capital letter, and the specific with a small letter (though it often incorrectly receives a capital one). Similar genera are grouped together into a 'family'. All manner of hierarchical groupings exist to link up the members of the plant kingdom but these are of little importance to the practical gardener. The classification schemes would be very neat and tidy were it not for the difficulty of defining exactly what constitutes a species. It cannot be the possession of a unique identity because much variation often exists even within what is generally accepted as a single species. Nor is the ability to cross-breed with other members of the same species but not with other species an acceptable criterion for the separation of plant species. Some botanists prefer to split off each distinct type of plant as a new species. Others like to keep the species as a much looser group within which considerable variation is possible. This is not completely satisfactory, however, and several subdivisions of a species are now accepted, one, of special significance to the gardener, is the cultivar, which is short for cultivated variety. The cultivar name, often also referred to as the variety, is used for a distinct form which has arisen in cultivation and possesses certain characters which enable it to be clearly distinguished from other forms of the same species. Those who are particularly interested in these aspects of naming plants are referred to the relevant International Codes (see bibliography).

THE INHERITANCE OF CHARACTERS

What concerns us in the present chapter is not the naming of plants but the creation of new forms. Several times, in earlier chapters, the question of characters which can be inherited has been mentioned. It is only very recently that botanists have become aware of exactly what the actual material of inheritance really is. The distinct form of any plant is due to the possession of certain characters. Some characters are 'qualitative', which implies an easily distinguishable distinction, such as the presence or absence of hairs. Others may be 'quantitative' with a difference only in, say, the number of flowers or the weight of fruit. These 'visible' characters are the result of 'factors' or *genes* present in the cells which make up the plant. It is now known that nearly all the genes are contained in the cell nucleus. The genes are the units of inheritance. In the simplest case each plant has two genes for each character, one coming from the pollen parent and the other from the female parent. These genes can exist in two, or even more, different forms. One form, for example, might give a tall pea plant and another produce a short one. If the two different forms of a gene occur together in one plant we can sometimes only see the effect of one of them. In our example the pea plant cannot be tall and short at the same time.

The idea that there exists a separate unit for each character in a plant resulted largely from breeding work of the type which Mendel and others had done. Gregor Mendel was an Augustinian monk who carried out his experiment in the 1860s. He found that certain characters in peas, such as the tallness and shortness already mentioned, were inherited in a very simple way. When, for example, a plant selected to breed true for tallness was crossed with a true-breeding short plant all the offspring were tall. (When we say that a plant breeds true we mean that all the offspring, in every generation, resemble the parents.) If we cross two distinct plants together the seeds which result and the plants which grow from them are known as the F_1 hybrid generation. In the most favourable cases these plants contain a mixture of the best characters from the two parents. For a few plants it is even possible to buy F_1 hybrid seed which has been produced for us by experts. Mendel found that his F_1 plants, when allowed to self-pollinate, gave approximately three tall plants for every short plant. These plants are called the F_2 generation. Obviously the F_1 generation contained a gene which could produce short plants in addition to the gene for the visible tall character. In complex cases, such as the cross between a single and a double dahlia, every one of the F_1 plants may be different. An F_1 hybrid is said to be *heterozygous*, that is, it possesses one of each type of gene by which its parents differed. In simple cases only one gene of a pair is 'expressed' in the F_1 and this is called the *dominant* one. In the example being discussed it is the tallness gene which is dominant. The 'hidden' character is due to a *recessive* gene.

For convenience the genes for a given character are usually represented by letters and where two characters affect the same feature the dominant form is given the capital letter and the recessive one the small letter. Assuming that the two genes of a pair pass independently to the offspring, we have the basis of a simple mathematical way of predicting the results of cross-breeding. The diagram overleaf shows a $3:1$ ratio in a simple cross, but this does not mean that if a single Tt plant produces only four seeds then three of them will always give tall plants and one a short plant. The ratio is predicted 'on the average' and dozens of plants are usually needed before figures which approach closely to the $3:1$ ratio are obtained.

When two distinct characters are considered in the parents, we arrive, by the same reasoning, to a simple *dihybrid ratio* of $9:3:3:1$ which means that there are four distinct types of offspring of which (on the average) nine will show both dominant characters, one will show both recessive characters and the re-

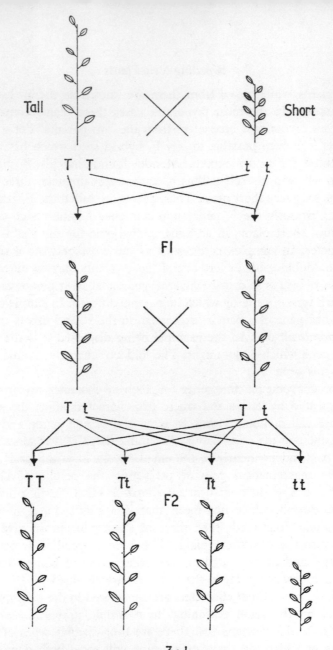

Tall   ✕   **Short**

T   T   t   t

FI

✕

T   t   T   t

T T   Tt   Tt   tt

F2

3 : 1

A representation of a cross between a tall and a short plant.
T is the dominant gene for tallness and t the recessive form
of the same gene which gives shortness

maining six will be equally divided with one of the dominant and one of the recessive characters each. All this may sound very easy but it soon becomes complicated when actual practical examples are considered. Sometimes the F_1 generation appears intermediate between the parents. When a white snapdragon is crossed with a red one the F_1 generation may all be pink and the F_2 generation consist of 1 red : 2 pink : 1 white. This is said to be due to *incomplete dominance* and the heterozygous individuals can easily be distinguished from the homozygous red or white ones. *Homozygous* means that the two genes present for a given character are of an identical form. Because the pink-flowered plants are heterozygous they will not breed completely true but will always give some red and some white flowered plants. In the case of 'quantitative' characters there are often several different 'additive' genes involved and the final appearance of a plant will depend upon just how many of these it posseses. In other cases different genes may interact together to upset the simple ratios. As a result of all this a large part of genetics (the study of heredity) is esentially a mathematical and statistical science. Such matters I leave to other books.

CHROMOSOMES

Almost all the genes responsible for the characters of a plant are carried on the chromosomes, which are only clearly visible at certain times. The cell nucleus has already been mentioned. At most times this appears merely as a distinct, roundish, spot when a cell is examined in a light microscope. Special stains help to reveal it clearly. When a cell is about to divide into two the contents of the nucleus become reorganized as thread-like structures. Again this is usually revealed with the help of special stains. At first the threads are rather indistinct and all tangled up but eventually they become clearly visible and much shorter. At this stage they are obviously split along their length. They then congregate

193

in one region, usually near the centre of the cell and each chromosome separates into two so that there is now twice the normal number of chromosomes. The two sets of chromosomes then move away from one another and turn back into 'resting' nuclei. A new cell wall then forms between them. Since nearly all the 'information' of a cell is contained in its nucleus this elaborate procedure ensures that, in the absence of some abnormality, every cell in a plant contains the same number of chromosomes and with it the ability to develop, under suitable conditions, into a complete plant. At first much of the information in the chromosomes is not used and it requires the processes of growth, differentiation and interaction with the world outside the cell before many of the characters are expressed.

Usually the various species of plant have sets of chromosomes which are visibly (with the help of a microscope and stains) distinct from one another. There may be a characteristic number of chromosomes, and the chromosomes may have a particular shape or size. An expert can tell quite a lot about a plant simply from looking at the chromosomes. Usually he will take a photograph of the chromosomes using the microscope and cut out the individual chromosomes from an enlarged print. He can then line them up neatly in rows and match them one with another. In a simple case he will find two of each sort. This is the state known as *diploid*. One chromosome in each pair derives from the pollen parent and the other one from the female parent, although it is impossible to tell visually which is which. There is a problem here which should now be obvious. Fertilization involves the fusion of a male and a female nucleus to give a single nucleus. If the egg cell and the fertilizing nucleus from the pollen had the normal cell number of chromosomes every fertilization would double up the number of chromosomes and the increase would soon get out of hand. So there is a special process by which the nuclei that join each other during fertilization have their chromosome number reduced to half the normal value.

When, as in a diploid nucleus, there are two chromosomes of each sort present it is only necessary to match up each pair and then cause the two sets to pass to opposite ends of the cell where they become 'reduced' nuclei (see photographs on page 136).

There are many plants which do not have the diploid number of chromosomes. If the chromosomes cannot separate into two equal sets the result is an upset in the 'reduction' division and the pollen or egg cells which form will have abnormal sets of chromosomes. This results in the plant not being very fertile or even in making it completely sterile, which in wild plants reproducing by seeds could result in the species dying out. In the garden vegetative means of propagation are used to perpetuate these unusual sets of chromosomes. It will be remembered that in ordinary cell division there is no pairing up of the chromosomes but simply a splitting of each one present so that it does not matter if there is no partner for it.

Many of the most important fruits and flowers are *triploids,* with three of each chromosome. These can arise if one of the parents produces its pollen or egg cells without the normal reduction division. Some of the best tulips and hyacinths are triploid. Bramley's Seedling and Blenheim Orange are just two examples of the many triploid apples. Because of the fertility problems mentioned above triploids are almost invariably propagated by vegetative means which is the only way of being sure of maintaining their characteristics. There are many plants which are even more complicated with four (*tetraploid*), five (*pentaploid*), six (*hexaploid*) and even more sets of chromosomes. The general term for all these is *polyploid*. By suitable methods it is even possible to obtain plants which have only one set of chromosomes (*haploids*). It is not necessary for a plant to have only multiples of complete sets of chromosomes and some have one or two extra or missing from some of the sets.

Quite early in the study of plant breeding it was discovered that some characters (genes) nearly always passed together to the offspring: they were 'linked' together. It can be shown that genes which show *linkage* are usually carried close together on the same chromosomes but it is possible for genes to cross over from one chromosome to another during the production of pollen or egg cells. Because of this the plant breeder is able to move particular genes about from one plant to another, although this is by no means as simple as it sounds. Normally only closely related plants, which have almost the same chromosomes, can be crossed together to give fertile offspring. Successful crosses between different genera are rare although there are several famous examples, such as *Fatshedera* produced from *Fatsia* and *Hedera* and *Raphanobrassica* from *Raphanus* and *Brassica*. When a cross is attempted between two different genera there are usually no viable seeds produced.

It is quite possible to carry out plant breeding without any knowledge of what exactly a gene or a chromosome is but, as with most things, the chances of success are much higher if one knows what one is doing. The chromosomes are partly made up of a special substance known as DNA, which is an abbreviation of its chemical name. The information in each of the genes which eventually results in the development of a particular character is now thought to be related to the exact order in which the subunits of the DNA molecules are joined together. The information in the DNA of a gene is first *transcribed* into a molecule of *'messenger* RNA'. This messenger can move out from the nucleus and into the tiny, but complex, structures which are concerned with the production of proteins. It will be recalled that all living reactions need the presence of protein catalysts, known as enzymes, before they can take place. A pigment, or any other character, cannot be produced unless the correct enzymes are present. The 'correct' enzyme is made up of amino acids joined together in a particular order. One of the great recent advances has been to relate the order of the four different subunits of DNA and RNA to the twenty or more different amino acids which can occur in proteins. Three of the subunits of RNA are equivalent to one amino acid and the *genetic code* lists the relationship between them. If the genes (DNA) are changed in any way the 'messenger' is different and different amino acids will be incorporated into the enzymes. This will cause a change in the characters appearing in the plant.

MUTATIONS

This may all sound very academic and distant from the garden but it is essential to the understanding of plant breeding. If the DNA of a gene is changed there may well be a sudden appearance of a new character—a 'sport' to the gardener, and to the botanist a gene *mutation*. Many of the interesting new plants which appear in gardens are sports. The Russell lupins, the Spencer-type tweet peas and many other valuable garden plants all arose in this way. Unfortunately the rate at which new mutations appear in the garden is very low. In any case most changes are detrimental rather than beneficial because a living system is so complex that it is much easier to upset it than to make it produce some interesting new character. The botanist gets over the problem of a low mutation rate by speeding it up artificially. Atomic radiation will cause mutations. So will many chemicals which react with the genetic material and change the information in it. The techniques are still very much a question of 'hit and miss' however because the way in which the characters are changed cannot yet be controlled predictably. After treatments to produce mutation most plants have undesirable characters, such as the absence of chlorophyll, or are unable to produce any viable offspring. Most of the effort goes into sorting out the few good mutations from the many bad ones and then manipulating the good ones to give a desirable cultivar for propagation and distribution. Unless he has received a very good biological and chemical education the gardener is well advised to stay away from the artificial induction of mutation: the methods used are so good at creating mutations that they could affect the careless gardener as well as his plants. So for 'sports' the gardener relies on natural ones or those supplied as new cultivars by professional growers who have been lucky enough to find them or have, in their turn, obtained them from someone else. Even the professional grower of garden plants does not usually risk trying to induce gene mutations artificially.

BREEDING TECHNIQUES

Most commercial and amateur plant-breeding work is concerned with rearranging existing genes. In this way recessive characters which were masked by a dominant gene can be revealed. New combinations of genes may give rise to new visible characters. Genes can also be introduced from related plants, which is one of the reasons why plant breeders like to have collections of plants from all over the world.

The first stage in any breeding programme is to select the parents. These could be just any two plants in the garden. As mentioned previously the chances of success are much greater if the plants are closely related and preferably of the same species. Those who want to breed new plants seriously will want to know, as far as possible, the pedigree of the parent plants. They will also need to keep detailed records of exactly what they do. For different species there are slightly different techniques which must be used to ensure cross-pollination. The essential thing at the start of a breeding programme is to make sure that only the selected pollen reaches the stigmas of the seed parent. This usually means removing the flower's own stamens before they liberate their pollen and covering the flower as soon as the desired pollen has been transferred (often by means of a small, soft paintbrush) to prevent unwanted pollen reaching the stigmas. Each hand-pollinated flower must be carefully labelled and its seeds kept separate from all others when they have formed. If the flowers to be crossed are not open at the same time it is sometimes possible to keep the pollen from one until the seed parent is ready but the exact conditions in which to store the pollen for more than a day or so are known for only a few plants.

Once the cross has been carried out all then depends on obtaining viable seed and germinating this to produce the F_1 plants. The fact that the offspring do not show any new desirable characters does not mean to say that none exist. They may merely be hidden. So this generation also has to be protected from random pollination and careful records kept of each plant. The seeds from each plant must be kept and labelled. So the breeding programme develops from year to year. As the numbers get too great to handle some will have to be discarded and only those plants which look promising grown on from year to year. A breeding programme is also an exercise in detailed book-keeping. Just as there are few people who win a fortune by betting so there are few gardeners who have developed a new plant without a great deal of effort. With slow-growing perennials, of course, one does not breed on for several generations. If the offspring when mature do not produce desirable flowers or fruit they are usually scrapped as also are any obviously weak or disease-prone seedlings, long before they reach maturity. Once a desirable perennial is found, vegetative propagation can be used to produce large numbers in a short time.

There are two different *breeding systems* in plants and this may have a marked effect on the outcome of any breeding programme. Actually the division is not absolute because some species show a mixture of both types. *Outbreeders* normally show cross-pollination and if they are continually forced to self-pollinate the vigour of the plants declines and yield, size and so on decrease. Many perennial plants are of this type. *Inbreeders* are normally self-pollinated and do not show any decline in vigour when forced to self-pollinate. Most annuals are of this type, which is fortunate, because if they were not it would prove difficult to obtain a new cultivar. Any new characters would be continually mixing with those from other plants due to cross-fertilization. One interesting thing about F₁ hybrids is that their parents are often not themselves particularly good. Nor are the offspring of F₁ hybrids the same as their parents. Whole books have been written on the subject of *hybrid vigour* or heterosis as it is known to the scientist.

Where the plants which are being cross-bred do not have the same chromosomes, or have unusual sets, then it may prove diffi-

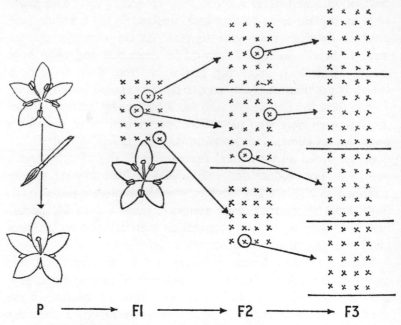

P ⟶ FI ⟶ F2 ⟶ F3

A diagram to illustrate a plant breeding experiment. It is assumed that only 20 seeds are produced per plant and only a few are selected for further breeding in each generation. If all the seeds were kept and grown there would be 8,000 plants in the F₃ generation

cult to obtain a viable offspring. The number of chromosomes and a great deal of information about the genetic make-up of cultivated plants is available for those who care to seek it out. For the serious plant breeder this information will often save a lot of time and wasted effort. Many of the common garden plants have very complicated pedigrees, often incompletely known because records were not kept of the early stages. The red currant, for example, contains some features from at least three different species of *Ribes*. Strawberries have a similar complicated history. Because of the contribution of several species to some cultivated plants it is often impossible to say exactly which 'species' forms the main part of the cultivated form. The strawberry, for example, is often referred to simply by the generic name and a cultivar name so that the gardener may sometimes meet *Fragaria* cv. 'Cambridge Favourite' which may at first sight appear botanically incorrect but is probably preferable to falsely attributing it to one species of *Fragaria*.

FORMATION OF POLYPLOIDS

Polyploids sometimes arise spontaneously or as a result of damage of one sort or another to a plant. It is easy to imagine that if the process of cell division could be stopped in the middle, after the chromosomes had split into two but before the new cell wall had formed then the cell would have twice its normal number of chromosome and be a tetraploid. Actually quite a lot of cells in a plant are like this but it usually takes some stimulus, such as decapitation, to cause a new tetraploid shoot to grow out. This can then be cut off and used as a cutting for propagation. So a new tetraploid plant will be obtained. Often, but by no means always, tetraploid plants are larger than the related diploid plant. They are rarely as fertile in producing seeds, however, because the process of reduction division may be upset by the presence of four chromosomes which are the same instead of

N

the usual two. There are many garden plants where this does not matter. Many years ago botanists discovered that the substance colchicine, which comes from the autumn crocus, was very effective in producing a doubling of the chromosome number. Most success is obtained with plants which have soft tissues and grow easily. Germinating seeds or young shoots can be soaked in a dilute solution of colchicine or a few drops of the solution can be put onto a growing apex. Often sterile hybrids can be made fertile by doubling up the chromosomes because this then gives two of each type to pair up before reduction division. It would be a mistake to think that every regrowth after damage is polyploid. More often it is perfectly normal.

SOMATIC MUTATIONS

Many variegated plants show interesting features related not only to the presence of changes in the cells but also to the way in which the different tissues of a plant are produced in the apex. If a change occurs in the cells which give rise to the central tissues of the stem and leaves they, but not the other tissues, will be abnormal. Such changes, as well as the polyploid cells referred to earlier, are examples of *somatic mutation*. In the variegated leaves there is a change in the chloroplast-containing cells which prevents them becoming green. Cells derived from unaffected parts of the meristem will be green and show through clearly to give the characteristic variegation pattern. It should be noted that there are many reasons for variegation and not all arise in the way which has just been described. The thornless blackberry 'Cory Thornless' arose by a loss of the 'thorn' character from the epidermal layers which cover the shoots. The cells underneath still have the thorn character, and seedlings, because they are not derived from the surface layers, all have thorns.

Plants in which some of the tissues have a different genetical constitution from the rest are called *chimeras*. Many different types of chimeras are recognized. In a *periclinal* chimera the inner tissues are surrounded by a layer or layers of cells with a different genetical make-up. *Sectorial* chimeras have a sector of changed tissues which extend from the centre of the stem to the epidermis. The thornless blackberries and thornless loganberries are periclinal chimeras. Some of the most interesting chimeras are those where the different tissues are derived from two completely separate plants. If a graft is made between two plants in the normal way and then the area of fusion between stock and scion is cut across a callus regenerates followed by young shoots. Most of the shoots are normal and belong to either the stock or the scion. A few may be mixed with part stock and part scion tissues. These are called graft chimeras. One of the most famous of this type is *Laburno-cytisus adamii* produced over a hundred years ago by bud grafting *Cytisus purpureus* onto *Laburnum anagyroides*. The new plant has a core of common laburnum covered by an epidermal layer of *Cytisus*. As might be expected parts of such 'dual' plants often develop shoots or at least small parts of shoots which are of only one type, usually *Laburnum*. If these are propagated they are completely normal. The dual plant has characters which are somewhere between those of the two grafted species.

At one time it was thought that graft chimeras were due to a fusion of the nuclei of the two grafted plants so that the information in them was mixed. It is now known that the known examples of chimeras all have distinct cells of one type or the other and that mingling of the chromosomes does not occur. Perhaps the best way of finishing this book is to note that botanists are now perfecting a startling new technique which could revolutionize plant breeding. Plant cells have been taken from two different plants and the cell walls dissolved away to leave the still-living cell contents as an unprotected *protoplast*. By suitable treatment it is possible to make the two protoplasts join together to give one mixed cell. If this can be made to regenerate into a whole plant—and this has already been done for normal cells—a totally new hybrid will be produced without the need to worry about pollination and fertilization, with all its consequent problems. It might even prove possible to mix together two totally unrelated plants in a way which was never thought possible before. The only point which limits one's enthusiasm is that previous experience with crosses between genera, using the ordinary methods of plant breeding, seem to indicate that it is not always the most desirable features of the parents which are expressed in the mixed offspring. Only time will tell us the value of these new techniques.

Bibliography

The following list is an entirely personal selection. Some of the books could well have been included under several chapters. Equally good books, which I do not mention, exist on many of the topics.

CHAPTER 1   HOW PLANTS GROW

Black, M., and Edelman, J. *Plant Growth*. Heinemann 1970
Burns, M. *Plant Anatomy*. The Arlington Practical Botany, Book 1. Arlington Books 1964
Duddington, C. L. *Evolution in Plant Design*. Faber 1969
Esau, K. *Plant Anatomy*. John Wiley 1967
Huxley, A. J. *Garden Terms Simplified*. David and Charles 1971
Leopold, A. C. *Plant Growth and Development*. McGraw-Hill 1964
Sinnott, E. W. *Plant Morphogenesis*. McGraw-Hill 1960

CHAPTER 2   TEMPERATURE, LIGHT AND WATER

Fish, M. *Gardening in the Shade*. David & Charles 1967
Hudson, J. P. 'The Plant and Its Environment', Parts I and II. *Journal of the RHS*, Vol XC III 1968
Kramer, P. J. *Plant and Soil Water Relationships*. McGraw-Hill 1969

Meidner, H. and Mansfield, T. A. *Physiology of Stomata.* McGraw-Hill 1969

North, F. *Perennials in a Cold Climate.* Gifford 1967

Smith, G. D. *Easy Plants for Difficult Places.* Collingridge 1967

CHAPTER 3
MINERAL NUTRITION AND OTHER FACTORS

Anderson, E. B. *Gardening on Chalk and Limestone.* Collingridge 1965

Buckman, H. O. & Brady, N. C. *The Nature and Properties of Soils.* Macmillan 1969

Carr, D. J. 'Chemical Influences of the Environment', *Encyclopaedia of Plant Physiology*, Vol XVI, pp 737-94. 1961

Cook, J. G. *Your Guide to the Soil.* Merrow 1965

Fish, M. *Gardening on Clay and Lime.* David & Charles 1970

HMSO Bulletin No 71. *Soils and Manures for Vegetables.* 1968

Hewitt, E. J. *Sand and Water Methods used in the Study of Plant Nutrition.* Commonwealth Agricultural Bureaux Technical Communion No 22 1966

Hollis, H. F. *Profitable Growing Without Soil.* English Universities Press 1964

Kelway, C. *Gardening on Sand.* David & Charles 1965

Kelway, C. *Gardening on the Coast.* David & Charles 1970

Russell, E. J. *The World of the Soil.* Collins 1957

Russell, E. W. *Soil Conditions and Plant Growth.* Longmans, Green 1961

Schutte, K. M. *The Biology of the Trace Elements.* Crosby Lockwood 1964

Small, J. *pH and Plants.* Bailliere, Tindall & Cox 1946

Stiles, W. *Trace Elements in Plants.* Cambridge University Press 1961

Treshow, M. *Environment and Plant Response.* McGraw-Hill 1970

Bibliography

Wallace, T. *The Diagnosis of Mineral Deficiencies in Plants by Visual Symptoms*. HMSO 1951

CHAPTER 4 CONTROLLING PLANT GROWTH

Cathey, H. M. 'The Growth Regulators Part 5. Ways of Applying Plant Retardants', *The Grower*, 8 March 1969
Evans, L. T. *Environmental Control of Plant Growth*. Academic Press 1963
Fraser, M. *The Gardener's Guide to Pruning*. Collingridge 1966
Hughes, H. M. 'Modern Techniques in Fruit Growing', *Journal of the RHS*, Vol XCVI(5) May 1971
Osborn, A. & Bagenal, N. B. *Pruning—Ornamental and Flowering Trees and Shrubs, Rose Trees and Fruit Trees*. Ward Lock 1962
Overbeek, J. van. 'The Control of Plant Growth', *Scientific American*, Vol 219(1), pp 75-81, July 1968
Roger, A. 'Bonsai. The Story and Development of the Dwarfed Tree', *Journal of the RHS*, Vol XCVI (6), June 1971
Shewell-Cooper, W. E. *The ABC of Pruning*. English Universities Press 1963
Wareing, P. F. 'The Physiology of Conifers', Parts I and II, *Journal of the RHS*, Vol XCVI (9) and (10) 1971
Wareing, P. F. & Phillips, I. D. J. *The Control of Growth and Differentiation in Plants*. Pergamon Press 1970
Yashiroda, K. *Bonsai, Japanese Miniature Trees*. Faber 1960

CHAPTER 5 SEEDS AND SEEDLINGS

Bunt, A. C. 'Peat-Sand composts: Their Value in Raising and Growing Ornamental Plants. I, General Principles', *Journal of the RHS*, Vol XCVI (1) Jan 1971
Crocker, W. *Growth of Plants*. Reinhold Publishing Co (N. York) 1948

Cooling, M. A. 'Peat-Sand Composts. Part II. The Use of Peat-Sand Composts in the Production of High Quality Bedding Plants', *Journal of the RHS*, Vol XCVI (2), Feb 1971

Hardwicke, G. D. *Flowers from Seed*. Barker 1967

Harrington, G. T. 'Optimum Temperatures for Flower Seed Germination', *The Botanical Gazette*, Vol LXXII (6), pp 337-58, 1921

Lawrence, W. J. C. & Newell, J. *Seed and Potting Composts*. George Allen & Unwin 1962

Mayer, A. M. & Roljakoff-Mayber, A. *The Germination of Seeds*. Pergamon Press 1966

Witham Fogg, H. G. *Growing Pot Plants from Seed*. Faber 1959

CHAPTER 6   VEGETATIVE PROPAGATION

Garner, R. J. *The Grafter's Handbook*. Faber 1967

Hunt, P. *Perennial Flowers For Small Gardens*. Barker 1966

Oldale, A. & Oldale, P. *Propagating Out of Doors*. Collins 1971

Pearse, H. L. *Growth Substances and Their Practical Importance in Horticulture*. Commonwealth Bureau of Horticulture and Plantation Crops, Technical Communication No 20, 1948

Thompson, P. A. 'Meristem Culture and Its Implications for Horticulture', *Journal of the RHS*, Vol XCVI (6), June 1971

CHAPTER 7   FLOWERING AND FRUITING

Blackwall, L. 'Pod-setting and yield in the runner bean (*Phaseolus multiflorus*)', *Journal of the RHS*, Vol XCVI (3), March 1971

Hillman, W. S. *The Physiology of Flowering*. Holt, Rinehart & Winston 1962

Hughes, A. P. & Cockshull, K. E. 'Effects of Night-break Lighting on Bedding Plants', *Experimental Horticulture No 16*, 1967

Lang, A. & Reinhard, E. 'Gibberellins and Flower Formation', *Advances in Chemistry*, 28, pp 71-9, 1961

Bibliography

Salisbury, F. B. *The Flowering Process*. Pergamon Press 1963
Vince, D. 'The Control of Flowering and Other Responses to Daylength', Parts I & II, *Journal of the RHS*, Vol XCV (5 and 6), 1970

CHAPTER 8   WEEDS, PESTS AND DISEASES

Agrios, G. N. *Plant Pathology*. Academic Press 1969
De Bach, P. *Biological Control of Insects, Pests and Weeds*. Chapman & Hall 1964
Duddington, C. L. *The Friendly Fungi*. Faber 1937
Fish, M. *Ground Cover Plants*. David & Charles 1970
Fryer, J. D. & Evans, S. A. *Weed Control Handbook*, Vols 1 and 2, Blackwell Scientific Publications 1968
Gram, E. & Weber, A. *Plant Diseases*. Macdonald & Co 1952
Harris, K. M. 'Notes from Wisley. A New Approach to Pest Control in the Dahlia Trial at Wisley', *Journal of the RHS*, Vol XCVI (5), May 1971
Hellyer, A. G. L. *Garden Pests and Diseases*. Collingridge 1966
Hunter, B. T. *Gardening Without Poisons*. Hamish Hamilton 1965
Lawrence, W. J. C. *Soil Sterilization*. George Allen & Unwin 1956
Martin, H. *Insecticide and Fungicide Handbook*. Blackwell Scientific 1969
Ogilvie, L. *Diseases of Vegetables*, HMSO Bulletin No 123, 1969
Ordish, G. *Biological Methods in Crop Pest Control*. Constable 1967
Salisbury, E. *Weeds and Aliens*. (New Naturalist Series) Collins 1964
Smith, K. M. *A Textbook of Plant Virus Diseases*. Churchill 1957
Wallace, H. R. *The Biology of Plant Parasitic Nematodes*. Edward Arnold 1963

Wood, R. K. S. *Biological Problems arising from the Control of Pests and Diseases.* Symposia of the Institute of Biology No 9 1960

CHAPTER 9 GREENHOUSES, FRAMES AND CLOCHES

Canham, A. E. *Artificial Light in Horticulture.* Centrex Publishing Co (Eindhoven) 1966
Clapham, S. *The Heated Greenhouse.* Collingridge 1964
Dakers, J. S. *The Modern Greenhouse.* Cassell 1967
Flawn, L. N. *Gardening with Cloches.* Gifford 1967
Flawn, L. N. & Flawn, V. L. *Gardening under Glass.* Gifford 1971
Fogg, H. G. *The Small Greenhouse.* Barker 1967
Gardiner, G. F. *Greenhouse Gardening.* Leonard Hill 1967
Goold-Adams, D. *The Cool Greenhouse Today.* Faber 1969
Macself, A. J. *The Amateur's Greenhouse.* Collingridge 1967
Menage, R. *Introduction to Greenhouse Gardening.* Phoenix House 1964
Simons, A. J. *All About Greenhouses.* Gifford 1967
Walkden, G. B. *Cloche Cultivation.* Collingridge 1955
Walls, I. G. *Greenhouse Gardening.* Ward Lock 1970
Welch, H. F. *Mist Propagation and Automatic Watering.* Faber 1970
Whitehead, G. E. *Grow Fruit in Your Greenhouse.* Faber 1970
Wood, J. P. *Greenhouse Management.* Collingridge 1962

CHAPTER 10 BREEDING NEW PLANTS

Beaty, J. Y. *Plant Breeding for Everyone.* Blandford
Burns, G. W. *The Science of Genetics.* Collier-Macmillan 1969
Clevenger, S. 'Flower Pigments', *Scientific American,* June 1964
Crane, M. B. & Lawrence, W. J. C. *The Genetics of Garden Plants.* Macmillan 1952

Bibliography

Darlington, C. D. & Janaki Ammal, E. K. *Chromosome Atlas of Cultivated Plants*. George Allen & Unwin 1945

Gowen, J. W. *Heterosis*. Hafner 1964

Handbook on breeding ornamental plants. Brooklyn Botanic Record: Plants and Gardens No 30, 1959

Hayes, H. K., Immer, F. R. & Smith, D. C. *Methods of Plant Breeding*. McGraw-Hill 1955

The International Code of Botanical Nomenclature and *International Code of Nomenclature for Cultivated Plants*. International Bureau of Plant Taxonomy and Nomenclature 1969

Jackson, A. A. Chrysanthemum Breeding at Wye College. *Journal of the RHS*, Vol XCVI (1), Jan 1971

Lawrence, W. J. C. *Practical Plant Breeding*. G. Allen & Unwin 1965

Lawrence, W. J. C. *Plant Breeding*. Edward Arnold 1968

McQuown, F. R. *Plant Breeding for Gardeners*. Collingridge

Neilson-Jones, W. *Plant Chimeras*. Methuen 1969

Plowden, C. C. *A Manual of Plant Names*. George Allen & Unwin 1968

Smith, L. J. *Plant Breeders' Rights*. Plant Variety Rights Office, Murray House, Vandon Street, London SW1

Thomas, D. G. *Simple, Practical Hybridising for Beginners*. Gifford 1957

Williams, W. *Genetical Principles and Plant Breeding*. Blackwell Scientific 1964

GENERAL

Annual Reviews of Plant Physiology. Annual Reviews Inc

Annual Reviews of Phytopathology. Annual Reviews Inc

Hay, R. & Singe, P. *Dictionary of Garden Plants*. Michael Joseph and Ebury Press 1969

Hessayon, D. G. 'Avoiding the Hidden Dangers in the Garden', *Journal of the RHS*, Vol XCVI (8), Aug 1971

Bibliography

Horticultural Abstracts. (Quarterly since 1931)

McLean, R. C. & Ivimey-Cook, W. R. *Textbook of Theoretical Botany,* Vols 1-3, Longmans, Green & Co 1951, 1956, 1967

Smith, A. W. *A Gardener's Book of Plant Names.* Harper & Row 1963

RHS Dictionary of Gardening and its new *Supplement* 1956 and 1969

Thrower, P. *Encyclopedia of Gardening.* Collingridge 1962

Tortora, G. J., Cicero, D. R. & Parish, H. I. *Plant Form and Function.* Collier-Macmillan 1970

Whitehead, S. B. *Everyman's Encyclopaedia of Gardening.* Dent 1970

Index

Page numbers in bold refer to plates

Abscisic acid, 31, 33
Abscission, 31, **83**
Acer saccharinum, 100
Actinomorphic flower, 79
Adventitious roots, 111
Aegopodium podagraria, 78
After-ripening of seeds, 101
Ammonia, damage by, 73
Apex, flowering, 131; root, 17
Apical dominance, 75-6, 80;
    meristem, 28
Aphid, 20
Apples, 149-50, 195
Araucaria excelsa, 78
Ash, bonfire, 60-61
Asters, 85
ATP, 23, 27
Auxins, 30, 32
Available water, 48

B9, 85
Bacterial diseases, 171
Bananas, 149
Begonias, 35, 114, 124
Beneficial micro-organisms, 171-2
Bonfires, 60-61
Bonsai, 91
Boron, 26, 68
Brassica, 196
Breeding systems, 199
Buddleia, 77, 86
Buds, 16, 17; later, 75-6
Buffer, 57

Cacti, 43
Calcium, 26, 57-8, 64, 67
Cambium, 129-30
Carbohydrates, 22
Carbon dioxide, 22
Carnations, sleepiness in, 73
Carrot, 99
CCC, 85
Celery, 39
Cell membrane, 41
Cell, plant, 27-9
Celosia cristata, 92
Chelating, 55
Cherries, 128
Chicory, 39
Chilling of seeds, 102
Chimeras, 202-4
Chlorine, damage by, 73
Chlorophyll, 23-5
Chloroplast, 29, **65**
Chromosomes, 193-7, **136**
Chrysanthemums, 85
Clay, 46, 57-8
Climacteric, 149
Climatic orbit, 173-4
Cloches, 178
Coal gas, 73
Colchicine, 202
Coleus, 82
Compensation point, 175
Competition, 88-9; with weeds,
    159-60
Compost, 59-60
Conifer, 76; sensitivity to sulphur
    dioxide, 72

213

Index

Convallaria, 102
Convolvulus arvensis, 112
Cork cambium, 43
Correlation, 81, 147
Cross-pollination, 198-9
Cytisus, 203
Cytokinins, 31, 33

2,4-D, 161; 2,4-DB, 162
Dahlia, 81, 114
Daylength, 39, 114, 134
Day-neutral plants, 134
Dicotyledons, 15
Digitalis, 79
Dihybrid ratio, 191
Diploid, 194
Disbudding, 80-81
DNA, 196
Dominant gene, 191
Drainage, 48
Dust, damage caused by, 74
Dwarfing, by chemicals, 85-6; by culture, 90-91; natural, 89-90

Ecosystem, 163
Embryos, **118**
Endosperm, 93, 98
Enzyme, 26
Etiolation, 38-9
Ethylene, in ripening, 149-50; toxic effects of, 73
Ethylene chlorhydrin, 119

Fasciation, 91-2
Fatshedera, 196
Fatsia, 196
Fertilizers, 53-4, 69-70
Field bindweed, 112
Field capacity, 48
Fixation, of carbon dioxide, 23
Flame gun, 156-7
Fluorine, 72
Food chain, 163
Forsythia, 76, 86
Foxglove, 79
Fragaria, 201
Frames, 179
Freesia, 79
Free space, 40

Frost damage, 35
Fungal diseases, 168-70, **135**

Gaillardia, 120
Genes, 190
Genetic code, 197
Gibberellins, 30, 33, 82-5
Golden cultivars, 24
Ground elder, 78
Growth substances, 29-33

Habitats, 163
Hardening, 35
Hedera, 196
Helianthus, 82
Helianthus tuberosus, 114
Heleborus niger, 104
Heterozygous plants, 191
Hevea, 100
Hoeing, 49
Homozygous plants, 193
Hormones, 29-33
Hybrid vigour, 200
Hydroponics, 53

IAA, 30, 75-6
IBA, 121
Incomplete dominance, 193
Internode, 17
Iron, 26, 55-6, 67-8

Klinostat, 79

Laburno-cytisus adamii, 203
Laburnum, 203
Lamium purpureum, 78
Lateral root, 17
Layering, 120
Leaf, 'copper', 25; mesophyll, 44; primordia, 16; shape, 16; variegated, 24; water loss from, 43-4
Leaf-fall, 31, 43
Lenticels, 43
Light, artificial, 23, 184-7; colours in, 23; effects on growth, 37-40; energy in, 23; in germination, 104
Lily of the valley, 102, 112
Limiting factors, 175
Linaria, 79, 92

Long-day plants, 134
Lotus, 100

Magnolia, 76
Maize, 57
Maleic hydrazide, 85
Maple, 100
Marigolds, 85
MCPA, 161
MCPB, 162
Meristem culture, 125-6
Meristem cells, 28
Mesophyll, 44
Messenger RNA, 196
Mist propagation, 183-4
Mitochondria, 29
Monocotyledons, 15, 17
Monopodial branching, 76
Morphactins, 85
Mulches, 36, 49

NAA, 121
Natural gas, 73, 181
Necrosis, due to toxic gases, 71-2
Nigella damascena, 104
Node, 17
Nodules (root), 25
Nucleus, 29, 193-4
Nutrients, and pH, 55-7

Orchids, dry sepal disease, 73
Organelles, 29
Organic matter, 21
Osmosis, 41
Overcrowding, 88-9
Oxygen, 22, 27
Ozone, 72, 73

Palisade cells, 44
PAN (peroxyacetyl nitrate), 72
Parsley, 99
Parthenocarpic fruit, 146
Pea, 25
Peach, 103
Peloric flowers, 79, 92
Peperomia sandersii, 124
Permanent wilting point, 48
Pesticides, 165-6
pH, 54-8

Petiole, 17
Phloem, 18-21
Phlox, 120; *drummondii*, 92
Phosphon, 85
Photon, 23
Photoperiodic induction, 114, 138
Photosynthesis, 21-5, 175-6
Phototropism, 39
Phytochrome, 104, 137-8
Pinching out, 80-81
Pisum sativum, 82
Plumule, 98
Polarity, 121
Polygonatum multiflorum, 112
Polyploid, 195, 201-2
Poppy, 120
Potato, 22, 113-19
Primary producers, 163
Primordia, 16, **84**
Protoplast, 203
Pruning, 86-8, **83**

Raphanobrassica, 196
Raphanus, 196
Recessive genes, 191
Red deadnettle, 78
Respiration, 26-7, 175-6
Rhizome, position in soil, 78,
    111-12
Rhubarb, 39
Ribes, 201
Ribosomes, 29
Rice, 30
Root, 17; 44-5; hairs, 18; lateral, 18;
    nodules, 25; pressure, 50
Rooting of cuttings, 121-3
Rubber plant, 100
Runners, 111-12

Saintpaulia ionantha, 124
Salix, 91
Sand, 57
Sap, 20
Shade plants, 24
Short-day plants, 134
Silicon, 57
Smoke, bonfire, 60
Smog, 72-3
Soil, pH, 54-7; structure, 45-7;

temperature, 36; water, 47-50
Solomon's seal, 112
Sophora japonica, 130
Spongy mesophyll, 44
'Sport', 197
Starch, 21-2
Stomata, 41, 65
Stratification, 102
Strawberry, 138, 201
Succession, 153
Succulents, 43
Sulphur, 26, 63-4
Sulphur dioxide, 71-2
Sunflowers, 35
Sun plants, 24
Sympodial branching, 76
Syringa, 76

Tea, 100
Temperature, effect on flowering,
    139-41
Thea, 100
Thiocyanta, 119
Toadflax, 79
Toxic gases, 71-4
Trace elements, 26, 56

Transformation, of bud, 77
Transpiration stream, 50
Triploids, 195
Tubers, 113-19
2,4-D, 161
2,4-DB, 162
2,4,5-T, 161

Vascular system, 18-19
Vegetative embryoids, 126-7
Vernalization, 102
Viruses, 170-1

Water, available, 48; types of, 182-3;
    uptake by seeds, 96-7
Watering, 42, 48-9, 182-4
Weedkillers, 160-63
Willow, 91
Wind, effect on growth, 70-71

Xylem, 18-19, 40

Yellow-leaved plants, 24

Zinc, 26, 28
Zygomorphic flowers, 79